Molecular potential energy functions

Molecular potential energy functions

J. N. Murrell, S. Carter, S. C. Farantos, P. Huxley and A. J. C. Varandas

School of Chemistry and Molecular Sciences
The University of Sussex

A Wiley–Interscience Publication

JOHN WILEY & SONS
Chichester · New York · Brisbane · Toronto · Singapore

Library of Congress Cataloging in Publication Data:
Main entry under title:

Molecular potential energy functions.

'A Wiley–Interscience publication.'
Includes index.
1. Excited state chemistry. 2. Molecular dynamics.
I. Murrell, J. N. (John Norman)
QD461.5.M65 1984 541.2′2 84-11821

ISBN 0 471 90540 2

British Library Cataloguing in Publication Data:

Molecular potential energy functions.
1. Molecules 2. Potential, Theory of
I. Murrell, J. N.
541.2′8 QD461

ISBN 0 471 90540 2

Typeset by Mid-County Press, London
Printed by St. Edmundsbury Press Ltd.,
Bury St. Edmunds, Suffolk

Contents

Preface

Molecular potential energy functions—a research monograph

In this monograph we describe advances that have been made in recent years in obtaining explicit functions for the potential energy surfaces of small polyatomic molecules. These functions are needed for any dynamical study of molecules including the rates of simple chemical reactions, molecular beam scattering cross-sections and vibrational and rotational spectroscopy. The potential energy function is needed both for classical and quantum mechanical treatments of this dynamics.

Information about the potential energy surface has improved enormously in recent years, both from the analysis of experimental data and from *ab initio* calculations, but it is still a major task to gather this information into a functional representation which can be used for calculation. In general, different experiments give information about different parts of the surface, and a particular *ab initio* technique may have different levels of accuracy for different parts of the surface. It is only by using chemical judgement that one can combine this information to produce a satisfactory function for the whole surface. Moreover, there is often a conflict between the accuracy with which a function represents a surface and the simplicity of the function; the more mathematically elaborate is the function, the more demanding will be the dynamical calculations in which this function appears.

Our book is based mainly on a series of papers which have been written during the past 10 years. These papers have employed techniques drawn from quantum chemistry, molecular dynamics, gas kinetics and vibrational spectroscopy. We put considerable emphasis on the fact that the functions developed have in many cases been tested by the dynamical calculations to which we have referred. Indeed, our experience has been that without these it is very difficult to make a judgement on the accuracy of a function.

Although some features of the functional form which we have used in our work are very much our own favourites, we hope that the book is not too narrow in its outlook. Our aim has been to put the emphasis on strategy rather than specificity. In other words, we emphasize the general features that potential energy functions should possess rather than the particular functions which have been used by us. We do recognize, however, that our strategy has only been tested on small polyatomic molecules, almost entirely those with three or four atoms. Whilst these cover a very important area of chemistry, it is of course small in the total. We have to leave unanswered the question of how far this strategy will take us for larger systems because our optimism is not yet supported by the hard work which will be necessary to confirm it.

The book has been written for readers who already possess a general understanding of the concept of the molecular potential energy function and a knowledge of the techniques, both computational and experimental, by which such potentials are deduced. In the first four introductory chapters we stress only a few topics which are particularly relevant to the later analysis. The emphasis of our book is not so much the molecular potential as the functional representation of such potentials.

Chapter 1

The molecular potential energy surface

1.1 Introduction

There have been considerable advances in recent years in developing analytical functions for the potential energy surfaces of small molecules. The concept of the potential energy surface for molecules is defined by the Born–Oppenheimer approximation for the separation of electronic and nuclear motion. The mathematical basis for this approximation has been extensively studied and is well documented[1], and we give no details in this book.

The molecular potential energy surface is the potential energy that determines the motion of the nuclei. In the Born–Oppenheimer approximation the electrons adjust their positions instantaneously to follow any movement of the nuclei, so that the potential energy surface can equally be thought of as the potential for the movement of atoms within a molecule or atoms in collision with one another. This type of motion is called adiabatic: the dynamics of the system are associated with a single potential energy surface. There are some molecular processes in which the system makes transitions between different potential energy surfaces. These are called non-adiabatic processes and they arise from factors which are ignored in the Born–Oppenheimer approximation.

The importance of adiabatic processes for chemical reactions was first emphasized by London[2], who suggested that most chemical reactions are adiabatic. Thus for a reaction such as

$$\mathrm{F}(^2\mathrm{P}) + \mathrm{H}_2(^1\Sigma_g^+) \rightarrow \mathrm{HF}(^1\Sigma^+) + \mathrm{H}(^2\mathrm{S})$$

the probability of the reaction can, in principle, be calculated by considering the dynamics of the three atoms under the influence of a single potential energy which is a function of the relative positions of the three atoms.

The other important field of chemical physics that can be understood largely by adiabatic processes is spectroscopy. The bound states of molecules can be classified by assigning definite electronic, vibrational, rotational and nuclear spin states, and the total energy can be represented as a sum of electronic, vibrational, rotational and nuclear spin energies. There are important features of spectroscopy for which this is not so, but the above description is almost always a useful first approximation. In particular, the potential energy surface associated with an individual electronic state of a molecule determines the vibrational energy levels associated with that state.

One can acquire knowledge relating to the potential energy surface either by analysing the results from some experiment which involves dynamical processes of the atoms, such as spectroscopy, molecular-beam scattering or reaction kinetics, or one can calculate the potential energy by first principles from the Schrödinger equation. The first of these approaches runs into what is generally called the inversion problem. Given the potential energy function, there is a straightforward, although usually computationally difficult, path for determining the dynamics of interest. There is usually not, however, a known route for moving in the opposite direction, that is, from dynamical data to a potential. In most cases the potential must be deduced by trial and error: assume a potential and deduce the corresponding dynamics; compare these with the observed dynamics and make adjustments to the potential to minimize the difference.

Although in recent years there have been enormous improvements in the computational techniques for determining the electronic energy levels of molecules by solving the Schrödinger equation, it still requires a great deal of effort to obtain the accuracy which is necessary to give results which agree well with experimental observations. The potential energy surface associated with a specific electronic state is the electronic energy for that state (including the nuclear–nuclear repulsion energy) for all configurations of the nuclei. Hence to calculate the potential energy surface from the Schrödinger equation one must solve the equation many times; it must be solved for each of the nuclear configurations that is thought necessary to build up an adequate representation of the surface.

For example, the shape of a triatomic system is defined by three variables (the three internuclear distances, say) so that the potential energy surface will be obtained as a table of values of the potential on a three-dimensional grid of the variables. One might typically require about 10^3 points to realize an adequate coverage of the surface. The computational task clearly gets rapidly out of hand as the number of atoms increases. However, this *ab initio* approach is still likely to be valuable for regions of the surface of special importance as it can give information that is impossible to obtain by analysis of experimental results.

To calculate the outcome of dynamical experiments from a given potential energy surface it is necessary to solve the equations of motion at the quantum classical or semi-classical levels. We will say very little about this computational problem in this book as it is well documented elsewhere[3]. The point we emphasize is that to solve the equations it is generally necessary to have explicit analytical functions for the potential energy surface. For example, at the classical level one would solve Hamilton's equations for the time evolution of the coordinates, q, and their conjugate momenta, p. Such equations are solved by numerical integration of a trajectory in the many-dimensional space (q, p) and the techniques for this require that the potential and its derivatives are quickly evaluated at any point on the surface covered by the trajectory. Therefore, even if one has a table of values of the potential at discrete points on the surface, it would be necessary to interpolate between these to obtain the necessary data for random points on the surface.

For some regions of the potential energy surface the analytical form can be deduced by theory without the computational effort required for an accurate solution of the Schrödinger equation. For example, it can be shown by perturbation theory that for two inert gas atoms separated by a large distance R, the interatomic potential can be expressed in the form[4]

$$V(R) = -(C_6 R^{-6} + C_8 R^{-8} + C_{10} R^{-10} + \cdots) \qquad 1.1$$

where the C_n are coefficients that are characteristic of the atom pair. Computational techniques exist for calculating these coefficients[5].

For clusters of atoms which are separated by distances of the order of typical chemical bond lengths it is difficult to deduce from theory the functional form of the potential surface. Nevertheless, it has become clear in recent years that functional forms of the potential can be found which, on *ab initio* and empirical evidence, provide a fairly accurate representation of the potential surface. It is the problem of selecting these functions and determining the parameters which they contain which is the principal topic of this book.

1.2 Diatomic potentials

We complete this chapter by some brief comments about the potentials for diatomic systems. They can be brief because diatomic potentials are essentially a solved problem and they have been well documented[6]. There are two reasons for considering the problem solved. Firstly, diatomic potentials depend on only a single variable, the interatomic distance, and the function, $V(R)$, is therefore much simpler than the multi-variable functions of polyatomic systems. Secondly, techniques exist for carrying out the inversion of both scattering data[7] and spectroscopic data[8,9] to give the potential. We do not wish to imply that all diatomic potentials are known with high accuracy. This is far from the truth, and there are many pairs of atoms for which the potential is not known even for the lowest electronic (ground) state, and others which have been much studied, such as the inert gas diatomics, which are still controversial when tested by the highest accuracy of experimental measurement.

A word of warning: we shall be concerned with the actual potential for a definite number of atoms. A widely used concept is the potential for a set of atoms when these are interacting with other atoms in the medium. An example is the potential between a pair of sodium atoms in liquid sodium. The best sodium pair potential to reproduce the properties of liquid sodium is not a true pair potential because it contains the average of the many-body potentials (the potentials between clusters of three or more atoms) which also influence the liquid properties. Pair potentials of this kind should be referred to as effective potentials and they have no validity outside the medium for which they have been defined.

The two most widely known diatomic potentials are the Morse function:

$$V(R)=D_e\{\exp[-2\alpha(R-R_e)]-2\exp[-\alpha(R-R_e)]\} \tag{1.2}$$

and the Lennard–Jones function:

$$V(R)=4\varepsilon[(\sigma/R)^{12}-(\sigma/R)^6] \tag{1.3}$$

The Morse function is used for two atoms which form a chemical bond, for example the ground state of two hydrogen atoms, and the Lennard–Jones function for atoms which do not form a bond but for which a shallow van der Waals minimum exists at rather large internuclear separations (for example, two neon atoms).

At large values of R the Lennard–Jones function is asymptotic to the curve $-4\varepsilon\sigma^6/R^6$ and comparison with expression (1.1) shows that it agrees with the leading terms of the theoretically determined potential if $4\varepsilon\sigma^6$ is equated with the parameter C_6. The repulsive branch of the Lennard–Jones function varies as R^{-12} but there is no theoretical justification for this form; it was a matter of convenience in early studies of liquids and gases to have the repulsive function equal to the square of the attractive function. It is more customary in modern work to give the repulsive branch an exponential form and the simple function[10]

$$V(R)=A\exp(-kR)-C_6R^{-6} \tag{1.4}$$

is a reasonable representation of inert gas potentials. However, no such simple function can reproduce the most accurate experimental data, and in Section 5.1 we will examine more complicated functions which have this capability.

The Morse function does not reproduce the correct asymptotic form of the potential at large values of R (this can be represented by a power expansion in R^{-1} even for open-shell atoms), but it is an adequate representation of the potential in the region of the potential minimum. The parameter α of the Morse function can be determined from the infrared spectrum of the molecule. In the harmonic approximation the vibration frequency of a molecule AB is given by

$$\nu_0=\frac{1}{2\pi}\sqrt{\frac{f}{m}} \tag{1.5}$$

where m is the reduced mass:

$$m=\frac{m_A m_B}{m_A+m_B} \tag{1.6}$$

and f is the force constant:

$$f=\left(\frac{d^2v}{dR^2}\right)_{R_e} \tag{1.7}$$

For the Morse function (1.2), differentiation gives

$$f=2D_e\alpha^2 \tag{1.8}$$

Hence α can be determined if ν_0 and D_e are known.

For a harmonic oscillator (i.e. a parabolic potential) the vibrational energy levels obey the equation

$$E_v = (v+\tfrac{1}{2})h\nu_0, \quad v=0, 1, 2\ldots \qquad 1.9$$

where ν_0 is given by expression (1.5) and the energies are measured relative to the minimum of the potential. The first level $v=0$, has an energy $\tfrac{1}{2}h\nu_0$, the so-called zero-point energy, and subsequent levels are spaced equally by $h\nu_0$. However, the vibrational energy levels of real molecules become closer together as the energy approaches the dissociation limit.

Rotation–vibration spectral lines can be identified with transitions between rotation–vibration energy levels each of which can be classified by two quantum numbers: v, the vibrational quantum number already defined by equation (1.9), and J, the rotational quantum number ($J=0, 1, 2,\ldots$). It is found empirically that the energy of a (v, J) level can be expressed as a convergent double expansion in the variables $(v+\tfrac{1}{2})$ and $J(J+1)$. It is customary to express the coefficients of this expansion in wavenumbers (reciprocal wavelength in cm^{-1}) to give the so-called term-value $T(v, J)$ of the level. Using standard nomenclature[11], the most important terms are as follows:*

$$\begin{aligned}\frac{E(v, J)}{hc} = T(v, J) &= \omega_e(v+\tfrac{1}{2}) - \omega_e x_e(v+\tfrac{1}{2})^2 + \omega_e y_e(v+\tfrac{1}{2})^3 + \cdots \\ &+ [B_e - \alpha_e(v+\tfrac{1}{2}) + \cdots]J(J+1) \\ &+ [D_e + \beta_e(v+\tfrac{1}{2}) + \cdots]J^2(J+1)^2 + \cdots \end{aligned} \qquad 1.10$$

The most extensive compilation of the parameters occurring in this expression is by Huber and Herzberg[12].

In 1932 Dunham[13] derived, by perturbation theory, expressions for the rotation–vibration energy levels of a diatomic molecule in terms of derivatives of the potential at the equilibrium distance. The Taylor expansion of the potential about R_e has the form

$$\begin{aligned} V(R) = V(R_e) &+ \frac{1}{2!}\left(\frac{d^2V}{dR^2}\right)_{R_e}(R-R_e)^2 + \frac{1}{3!}\left(\frac{d^3V}{dR^3}\right)_{R_e}(R-R_e)^3 \\ &+ \frac{1}{4!}\left(\frac{d^4V}{dR^4}\right)_{R_e}(R-R_e)^4 + \cdots \end{aligned} \qquad 1.11$$

which can be written as

$$V(\rho) = V(0) + \tfrac{1}{2}f_2\rho^2 + \tfrac{1}{6}f_3\rho^3 + \tfrac{1}{24}f_4\rho^4 + \cdots \qquad 1.12$$

where $\rho = R - R_e$ is the displacement from equilibrium and f_2, f_3 and f_4 are the second, third and fourth derivatives, respectively, of the potential at $\rho=0$.

As was shown by Dunham, the derivatives of the potential can be obtained from those parameters in equation (1.10) which multiply the variable $(v+\tfrac{1}{2})$. The

* D_e appearing in expression (1.10) should not be confused with the dissociation energy which appears, for example, in the Morse potential (1.2).

second derivative is given by the same expression as for the harmonic oscillator. If we identify ν_0 with $c\omega_e$ and f with f_2 we have, from equation (1.5),

$$f_2 = 4\pi^2 m\omega_e^2 c^2 \qquad 1.13$$

The third derivative can be determined from α_e through the expression[13]

$$f_3 = \frac{-3f_2}{R_e}\left(1 + \frac{\alpha_e\omega_e}{6B_e^2}\right) \qquad 1.14$$

and the fourth derivative from $\omega_e x_e$ by

$$f_4 = \frac{f_2}{R_e^2}\left[15\left(1 + \frac{\alpha_e\omega_e}{6B_e^2}\right)^2 - \frac{8\omega_e x_e}{B_e}\right] \qquad 1.15$$

In principle, higher derivatives can be obtained from the higher-order terms in equation (1.12), but in practice it is rarely useful to go beyond f_4.

Table 1.1 gives the spectroscopic parameters for H_2 and CO and the resulting derivatives of the potential deduced from expressions (1.13)–(1.15). Figures 1.1 and 1.2 show the Taylor expansions to the potentials for these molecules, which have been terminated at the second, third and fourth derivatives, respectively. Comparison is made with points on the potential curve which have been deduced by the RKR (Rydberg–Klein–Rees) method[9,14].* It can be seen that the range over which these Taylor expansions are a good fit to the exact potential is very small; even taken to the fourth order the Taylor expansion is significantly in error for $|R - R_e| \approx 0.2$ Å.

Figure 1.3 compares the Morse potential (1.2) for H_2 with the RKR points. Although the Morse function only reproduces correctly f_2, the second derivative of the potential at the minimum, it is a much better overall fit to the true potential than the Taylor expansion to the fourth order. The key factor leading to this improvement is that the Morse function has the correct asymptotic behaviour as R approaches infinity. The point to be stressed is that no matter how much care is taken to fit the potential close to the minimum, if one does not impose the correct asymptotic behaviour at infinity the potential will be useless for studying atomic collisions, or even for high-energy rotation–vibration states of the molecule.

* The Rydberg–Klein–Rees method uses semi-classical mechanics to go directly from diatomic term values to a tabulated function of $V(R)$, without making any assumption about the functional form of $V(R)$. Semi-classical quantization of a periodic motion provides the condition

$$\oint P\,\mathrm{d}R = \oint [2\mu(E-U)]^{1/2}\,\mathrm{d}R = (v' + \tfrac{1}{2})h$$

where P is the momentum conjugate to R and the integration is over a complete cycle of the motion. U is the potential, including the centrifugal energy, and depends on the rotational quantum number J as follows:

$$U(R, J) = V(R) + J(J+1)\hbar^2/2\mu R^2$$

For a given (v, J) state one can deduce $R_{max} - R_{min}$, the difference between the two classical turning points in U at energy E above the minimum of the potential. As $R_{max} - R_{min}$ is a function of J as well as v, one can, by comparing different J states for a given v, determine both R_{max} and R_{min} for each v state. The RKR method can only be used for diatomic molecules.

Table 1.1 Dunham expansion parameters[12] and force constants for CO and H_2. aJ ≡ mdyne Å.

Parameter	CO	H_2
R_e/Å	1.128323	0.74144
ω_e/cm^{-1}	2169.81358	4401.213
$\omega_e x_e$/cm^{-1}	13.28831	121.336
$\omega_e y_e$/cm^{-1}	0.010511	0.8129
B_e/cm^{-1}	1.93128087	60.8530
α_e/cm^{-1}	0.01750441	3.0622
f_2/aJ Å^{-2}	19.017	5.751
f_3/aJ Å^{-3}	−136.4	−37.4
f_4/aJ Å^{-4}	808	238

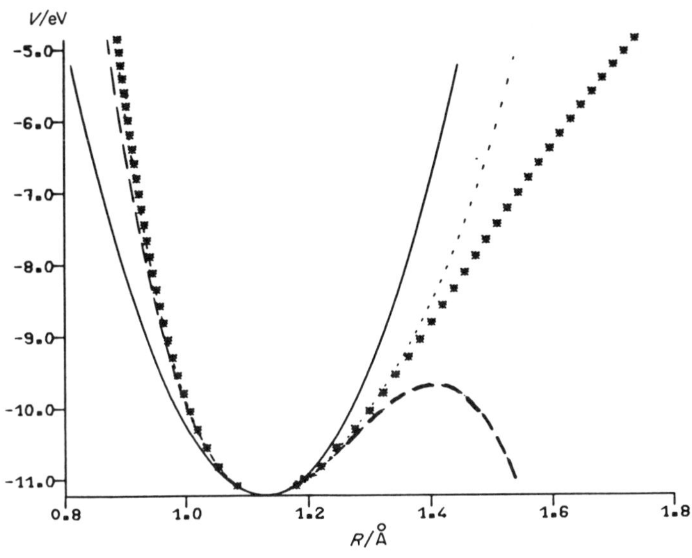

Figure 1.1 Comparison of the Taylor expansions to the CO potential terminated at the quadratic (——), cubic (— —) and quartic (- - - -) levels with the RKR values of the potential shown as stars

Starting with a function such as the Morse function, which has the correct asymptotic behaviour, and introducing more flexibility into the function, it is possible to improve the fit to the exact potential around the potential minimum. The Morse function (1.2) was extended in this way by Hulbert and Hirschfelder[15]. By multiplying the repulsive branch of the potential by a polynomial in $\rho = (R - R_e)$, as follows:

$$V_{HH} = D_e[(1 + g\rho^3 + h\rho^4)\exp(-2\alpha\rho) - 2\exp(-\alpha\rho)] \qquad 1.16$$

they were able, by a suitable choice of g and h, to fit the cubic and quartic (as well

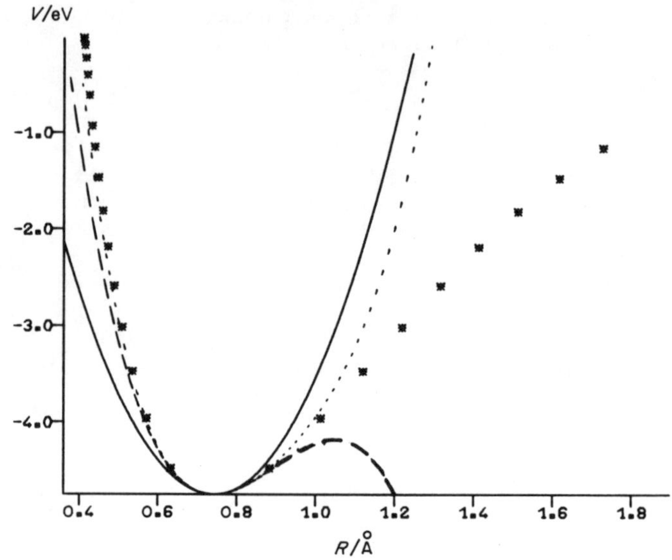

Figure 1.2 Comparison of the Taylor expansions to the H_2 potential with the RKR points. Symbols as in Figure 1.1

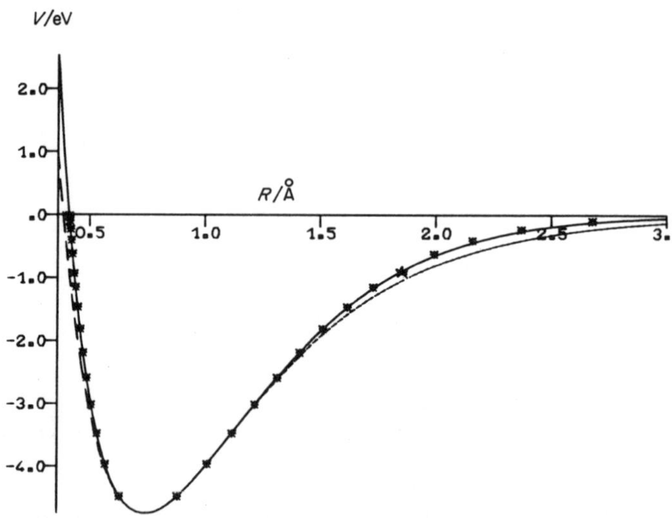

Figure 1.3 Comparison of the Morse function (1.2) and the Extended Rydberg function (1.19) with the RKR points for H_2. The two functions are shown as continuous curves, that of the ER function passing through all the RKR points. The Morse potential is too negative at large R

as the quadratic) derivatives of the potential at the minimum. However, as the attractive branch of the Morse potential is not modified by this procedure, the Hulbert–Hirschfelder function is not a great improvement over the simple Morse function for $\rho > 0$.

A more satisfactory function which can be extended to fit higher derivatives of the potential is a little used function first suggested by Rydberg[16]:

$$V_R = -D_e(1+a\rho)\exp(-a\rho) \qquad 1.17$$

This function has a minimum of depth D_e at $\rho=0$ ($R=R_e$), and the second derivative of the potential at $\rho=0$ is given by

$$f_2 = a^2 D_e \qquad 1.18$$

The Rydberg function is as good a representation of a typical diatomic potential as is the Morse function. Moreover, as was shown by Murrell and Sorbie[17], it is more easily extended to fit higher derivatives of the potential by simply increasing the order of the polynomial in equation (1.17).

The Extended Rydberg (ER) function is defined as

$$V_{ER} = -D_e\left(1+\sum_k a_k\rho^k\right)\exp(-\gamma\rho) \qquad 1.19$$

which has a minimum at $\rho=0$ ($R=R_e$) if $\gamma=a_1$. To fit f_2, f_3 and f_4 one requires terms in the polynomial to $k=3$. The relationships between the a_k and these derivatives are as follows:

$$f_2 = D_e(a_1^2 - 2a_2)$$
$$-f_3 = D_e(2a_1^3 - 6a_1a_2 + 6a_3)$$
$$f_4 = D_e(3a_1^4 - 12a_1^2a_2 + 24a_1a_3) \qquad 1.20$$

The coefficients a_2 and a_3 can be eliminated from these expressions to give a quartic polynomial in a_1:

$$D_e a_1^4 - 6f_2a_1^2 - 4f_3a_1 - f_4 = 0 \qquad 1.21$$

which has, for most values of the spectroscopic parameters, only one positive root. The equation can be solved for this root and, by substituting back into equations (1.20), a_2 and a_3 can be determined. If there is more than one positive root of equation (1.21), the largest will give the best potential.

Table 1.2 gives the parameters of the ER potential (1.19) for CO and H_2, which were deduced from the force constants of Table 1.1 and the dissociation energies. Figure 1.3 shows that the ER function for H_2 is in excellent agreement with the RKR points, and that for CO is no less accurate.

For almost all molecules that we have studied[18] the ER function has positive values for a_1, a_2 and a_3, and has only one stationary point, namely that at $\rho=0$. However, there is no general rule that this is the case and the functions should always be examined for spurious turning points in the physically important range of ρ.

The simplest and most widely used function for repulsive states is the so-called Born–Mayer function:

$$V(R) = A\exp(-bR) \qquad 1.22$$

Table 1.2 Parameters of the Extended Rydberg function for the ground-state potentials of CO and H_2.

Parameter	CO	H_2
$R_e/\text{Å}$	1.1283	0.7414
D_e/eV	11.226	4.7472
$a_1/\text{Å}^{-1}$	3.897	3.961
$a_2/\text{Å}^{-2}$	2.305	4.064
$a_3/\text{Å}^{-3}$	1.898	3.574
$\rho/\text{Å}^{-1}$	3.897	3.961

where A and b are constants. It can be seen that this is a special case of the ER function (1.19) in which the a_k are zero and D_e is not identified with a dissociation energy. Thus the ER function can be used to represent repulsive states by fitting the parameters to *ab initio* calculations of the potential. For example, the potential of the lowest triplet state of H_2 which has been calculated with high accuracy[19] is well represented in its repulsive region by the function

$$V\,(\text{eV}) = 5.8568(1 + 1.4782\rho + 1.7802\rho^2)\exp(-3.9022\rho) \qquad 1.23$$

where

$$\rho\,(\text{Å}) = R\,(\text{Å}) - 0.7417 \qquad 1.24$$

The ER function, like the Morse function, does not reproduce the dispersion energy of two atoms at large distances. To correct for this a term in R^{-6}, and perhaps higher powers of R^{-1}, can be added to the ER function. Alternatively, the range of R can be divided into two, an inner where the ER function applies and an outer where an inverse polynomial expansion applies[20]. Such functions will be discussed fully in Section 5.1.

Our main objective in giving this account of diatomic potentials was to prepare the ground for a description of polyatomic potentials. The main point which we have made is that to obtain a potential which is satisfactory over all physically accessible values of R it is necessary to impose the correct asymptotic behaviour of the potential for dissociation into atoms. If this is done, then rather simple potentials can be used which are nevertheless reasonably accurate in the region of the equilibrium distance. We have spent more time than might be expected on the Extended Rydberg function because it plays an important role in our treatment of polyatomic potentials, not only for their diatomic components but also because the ER function can be easily extended to greater dimensions. Before we turn to this problem, in the next chapter we give a general account of the topography of polyatomic surfaces.

References

1. M. Born and K.Huang, *Dynamical Theory of Crystal Lattices*, Oxford University Press, London, 1954.

2. F. London, *Z. Electrochem.*, **35**, 552 (1927).
3. M. S. Child, *Molecular Collision Theory*, Academic Press, London, 1976; W. H. Miller (Ed.), *Dynamics of Molecular Collisions*, Plenum Press, New York, 1976.
4. G. C. Maitland, M. Rigby, E. B. Smith and W. A. Wakeham, *Intermolecular Forces, Their Origin and Determination*, Clarendon Press, Oxford, 1981; M. L. Klein and J. A. Venables (Eds.), *Rare Gas Solids*, Academic Press, New York, 1976.
5. A. Dalgarno, *Adv. Chem. Phys.*, **12**, 143 (1967); G. Starkschall and R. G. Gordon, *J. Chem. Phys.*, **54**, 663 (1971).
6. J. Goodisman, *Diatomic Interaction Potential Theory*, Vols. 1 and 2, Academic Press, New York, 1973.
7. U. Buck, *Adv. Chem. Phys.*, **30**, 313 (1975); *Rev. Mod. Phys.*, **46**, 369 (1974).
8. R. J. LeRoy, in *Semi-Classical Methods in Molecular Scattering and Spectroscopy*, Ed. M. S. Child, Reidel, Dordrecht, 1980, Ch. 3.
9. W. G. Richards and R. F. Barrow, *Trans. Faraday Soc.*, **60**, 797 (1964); *Proc. Phys. Soc.*, **83**, 1045 (1964).
10. R. A. Buckingham, *Proc. Roy. Soc. London, Ser. A*, **168**, 264 (1938).
11. G. Herzberg, *Spectra of Diatomic Molecules*, Van Nostrand, Princeton, 1950.
12. K. P. Huber and G. Herzberg, *Constants of Diatomic Molecules*, Van Nostrand, Princeton, 1979.
13. J. L. Dunham, *Phys. Rev.*, **41**, 721 (1932).
14. J. T. Vanderslice, E. A. Mason, W. G. Maisch and E. R. Lippincott, *J. Mol. Spectrosc.*, **3**, 17 (1959); **5**, 83 (1960).
15. H. M. Hulbert and J. O. Hirschfelder, *J. Chem. Phys.*, **9**, 61 (1941).
16. R. Rydberg, *Z. Phys.*, **73**, 376 (1931).
17. J. N. Murrell and K. S. Sorbie, *J. Chem. Soc., Faraday Trans. 2*, **70**, 1552 (1974).
18. P. Huxley and J. N. Murrell, *J. Chem. Soc., Faraday Trans. 2*, **79**, 323 (1983).
19. W. Kolos and L. Wolniewicz, *J. Chem. Phys.*, **49**, 404 (1968).
20. A. J. C. Varandas, *J. Chem. Soc., Faraday Trans. 2*, **76**, 129 (1980)

Chapter 2

The topography of potential energy surfaces

2.1 Hills, valleys and other features

Most diatomic potential energy curves have a very simple form: either they have a single minimum corresponding to a bound state of the two atoms or they are repulsive with, generally, only a shallow van der Waals minimum at large internuclear distances. It is comparatively rare (very rare for ground states) to find curves with more than one stationary point and, when one does, this is usually due to an avoided crossing of two curves, a topic we examine in the next section.

The potential energy surfaces of polyatomic systems are more complex than diatomic potentials, not only because of their greater dimensionality but also because there are features that are specifically associated with this greater dimensionality. However, polyatomics and diatomics have in common that their potentials are generally smooth. A diatomic potential does not look like a cross-section through the Alps. Polyatomic surfaces are not jagged; a gentle countryside of rolling hills is a better image. This geographical image is important because it implies that the mathematical task of representing the surface by relatively simple functions may not be as daunting as some have assumed. Just as Morse or Rydberg functions are simple and fairly accurate representations of diatomic curves, so such functions, increased in dimension, may be adequate representations of polyatomic surfaces.

A triatomic molecule has a potential energy surface which is a function of three independent internal coordinates. A four-atom surface is six-dimensional, an n-atom surface is $3n-6$ dimensional. We shall consider later the question of what coordinates should be used to represent the potential.

The most important features of a surface are its stationary points, those places on the surface which have zero gradient, i.e. where $3n-6$ independent first derivatives are zero. These stationary points can be of several types, depending on the nature of their second derivatives. However, before exploring this in more detail we note that a minimum on the surface is simply defined as a point at which an infinitesimal step in any direction leads to an increase in potential energy. A polyatomic surface can have more than one minimum; in fact, real systems are more likely to have several minima than just a single minimum. The lowest point on the surface will represent the most stable arrangement of the nuclei for that particular electronic state and other minima on the surface represent metastable species.

For example, the ground-state surface of the triatomic system HCN has two minima, one representing the equilibrium geometry of the molecule HCN and the other the equilibrium geometry of the metastable molecule HNC. The lifetimes of metastable species can be short or long and will depend on external conditions. Very long-lived metastable species can be treated as normal molecules. For example, dimethyl ether, $(CH_3)_2O$, is actually a metastable species on the same surface as the more stable ethanol, C_2H_5OH.

Maxima on a surface can be defined as simply as minima: an infinitesimal step in any direction leads to a decrease in potential. However, maxima are not generally of special physical significance. Much more important topographical features are saddle points. In qualitative terms a saddle is a point which has one, but only one, direction in which the second derivative of the potential is negative, and along all other orthogonal directions it is positive. To be more precise we must examine the second derivatives in more detail.

In terms of the internal coordinates ρ_i, the second derivatives of the potential at a stationary point can be expressed as a matrix with elements

$$V''_{ij} = \frac{\partial^2 V}{\partial \rho_i \, \partial \rho_j} \qquad 2.1$$

By an orthogonal transformation of these coordinates it is possible to obtain new coordinates in which V'' is a diagonal matrix. These coordinates define the principal coordinates at the point. If all the diagonal elements are positive the point is a minimum on the surface. If one diagonal element is negative and all others are positive, the point is a saddle.

Saddle points are of particular interest in chemical kinetics because they lie on the paths between points on the surface identified with reactant molecules and points on the surface identified with product molecules. A saddle point is the *highest point* on the path which involves the *lowest* increase in the potential on passing from reactants to products. Stationary points also exist which have more than one negative curvature along the principal axes. However, these do not generally have any special kinetic significance because an 'easier' path between reactants and products (less increase in potential) can always be found by by-passing such points[1].

The dissociation products of a polyatomic system correspond to regions at infinity where the potential surface is flat in one or more dimension, i.e. all derivatives in these directions are zero. For example, the dissociation of a triatomic species ABC to A plus BC is associated with paths into a region where $R_{AB} = R_{AC} = \infty$ and all derivatives with respect to these variables become asymptotically zero. These asymptotic regions can be referred to as valleys. As A approaches BC so the slope of the valley gradually changes. If ABC is not a stable triatomic species (e.g. H_3), then the slope will be positive and the path along the bottom of the valley will move up to a saddle point (in this case the saddle represents a collinear species with two equal H—H bond lengths) and pass over down the reactant valley. If ABC is a stable species then the valley may simply dip down to the minimum or there may be a saddle point on the way.

Our references above to paths between reactants and products does not imply that, in a classical sense, the internal coordinates change according to a single pattern (defined by the path) during the chemical reaction. The correct classical analogue is that the reaction proceeds along a statistical distribution of paths or, as they are usually referred to, classical trajectories. Nevertheless, the special paths that pass from reactants to products along the bottom of valleys and through saddle points are of chemical interest because they feature in several qualitative theories of chemical dynamics. The distance along such a path is called the reaction coordinate and a graph of V along a reaction coordinate is called the potential energy profile for the reaction.

For example, in the dissociative reaction

$$\mathrm{ABC} \rightarrow \mathrm{AB} + \mathrm{C}$$

as the B—C bond is broken there will generally be small changes in the AB bond length and the angle $\mathrm{A\hat{B}C}$. Thus the reaction coordinate will, for the most part, be parallel to the internal coordinate R_{BC}, but not quite parallel, to allow for these other coordinate changes. In this example the reaction profile would be very similar to the potential energy curve for dissociation of a diatomic molecule and simple functions such as Morse or Rydberg functions would be expected to be a good representation of such a profile.

Although we have stressed the concept of smoothness for polyatomic potential surfaces, there is one complication which belies this term and that is associated with the crossing of surfaces. As this feature has important implications for the determination of functions to represent polyatomic surfaces we shall describe it in detail.

2.2 Surface crossing

An important step in our understanding of diatomic potential energy curves was the proof by von Neuman and Wigner in 1929 that the curves associated with states of the same spin multiplicity and spatial symmetry could not cross[2]. Those of different spin or spatial symmetry may cross. In recent years there have been a number of papers commenting on the rule and proofs of the rule (for a recent discussion, see ref. 3), but the situation has now settled down to essentially the early position, and the non-crossing rule can be accepted with few qualifications. The main one is that for one-electron homonuclear diatomic molecules there is a separation of variables in the wave function over and above those associated with symmetry criteria. In such cases the potential curves of two states which have the same symmetry, but are otherwise distinguishable in terms of these variables, can cross. The best-known example is the crossing of the $2s\sigma_g$ and $3d\sigma_g$ states of H_2^+ [4].

The proof we give of the non-crossing rule is based on that of Teller in 1937[5], because this proof provides the simplest method of extending the rule to polyatomic systems and it also introduces the important concepts of diabatic and adiabatic surfaces.

Let ψ_1 and ψ_2 be the wave functions of two electronic states which have the same symmetry and spin. We assume that these two wave functions can be written as a linear combination of two orthonormal basis functions ψ_a and ψ_b. In that case, the energies of the two states are the eigenvalues of the 2×2 Hamiltonian matrix

$$\begin{pmatrix} H_{aa} & H_{ab} \\ H_{ab} & H_{bb} \end{pmatrix} \tag{2.2}$$

which are given by

$$E = \tfrac{1}{2}\{H_{aa} + H_{bb} \pm [(H_{aa} - H_{bb})^2 + 4H_{ab}^2]^{1/2}\} \tag{2.3}$$

In real space these two energies will be equal only if

$$H_{aa} - H_{bb} = 0 \tag{2.4}$$

and

$$H_{ab} = 0 \tag{2.5}$$

If $H_{aa} - H_{bb}$ and H_{ab} are independent functions of the internuclear distance, R, then they will not, except by accident, be simultaneously zero.

If ψ_a and ψ_b have different symmetries then H_{ab} will be zero for all values of R. In that case there may be a point or points at which equation (2.4) is satisfied, i.e. the energies of the two states are equal. These points will then be crossing points of the potential energy curves. We refer the reader to other sources for the arguments for, and against, this simple proof[3,6].

Teller[5] was the first to point out that if $H_{aa} - H_{bb}$ and H_{ab} are independent functions of two variables then it is possible, in some situations, to satisfy both equations (2.4) and (2.5). More generally, if there are $3n-6$ internal coordinates then equations (2.4) and (2.5) could be satisfied in a space of $3n-8$ dimensions. Thus for a triatomic molecule the crossing of potential energy surfaces for states of the same symmetry can occur in one dimension, that is, along a line. This result was later amplified by Herzberg and Longuet-Higgins[7].

If in a triatomic system one internal dimension is kept constant then surface intersections can occur at a point in the remaining two-dimensional space. This is referred to as a conical intersection of two surfaces, and is illustrated schematically in Figure 2.1. This type of intersection has been extensively discussed for the Jahn–Teller situation in which the crossing point represents a structure belonging to a non-abelian symmetry group (i.e. a group for which degeneracies can occur).

An important criterion for identifying such points was given by Longuet-Higgins *et al.*[9] for the Jahn–Teller case, and later generalized by Longuet-Higgins[6] for all such intersections. On going once around any closed path on the

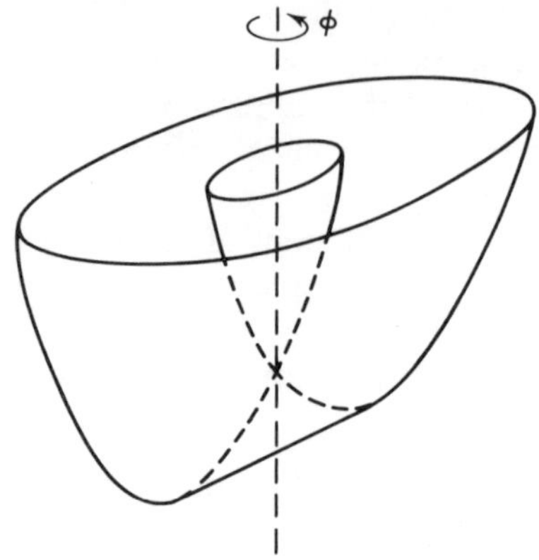

Figure 2.1 A conical intersection of two surfaces. In a typical Jahn–Teller situation the surfaces would show oscillations having three-fold symmetry about the cone axis[8]

surface that contains the conical intersection the sign of the electronic wave function will change. Using this criterion an intersection has been identified in the potential energy surfaces of the system LiNaK[10]. This crossing is not an attribute of symmetry as the system belongs to the C_s group which has no degenerate representations. As is well known, the simple three-electron system H_3 has a conical intersection at points with D_{3h} symmetry and in the full H_3 surface there is a line of such crossing points (the extra dimension being the perimeter of the three-atom triangle).

Teller's proof of the non-crossing rule provides an introduction to the concept of diabatic and adiabatic surfaces. The term 'adiabatic' is used for the exact surface defined within the Born–Oppenheimer approximation. It has the implication that if the dimensions of a molecular system change continuously in accord with some classical path on a potential energy surface, then the electronic wave function of the system will also change continuously. Most chemical reactions are adiabatic, that is, the electronic state of the system is at all times, from reactants to products, associated with one potential energy surface. Non-adiabatic reactions do occur under some conditions and they require the system to make a transition between two or more adiabatic surfaces.

In cases where a non-adiabatic process is highly probable the dynamics may be understood more simply by considering the motion on non-adiabatic surfaces, usually called diabatic surfaces[11]. Diabatic surfaces do not satisfy the Born–Oppenheimer approximation, indeed the diabatic electronic states may not be eigenfunctions of any defined electronic Hamiltonian. The states ψ_a and ψ_b referred to in the Teller proof of the non-crossing rule are diabatic states and the energies H_{aa} and H_{bb} which appear in the matrix (2.2) are diabatic potentials. There may be situations in which diabatic states are defined with respect to some approximate Hamiltonian, or are approximate wave functions for the system, and others in which they have no physical significance. Nevertheless, as we shall see in later chapters, they are important for the task of producing explicit functions for the potential energy surfaces when surface crossings occur of the type which we have been describing.

Although we have been writing about the intersection of surfaces, it can be seen from Figure 2.1 that when these intersections are of the conical type (i.e. associated with states of the same symmetry), strictly it is not possible to refer to separate surfaces. It is possible to pass between the lower and upper surfaces of the figure by passing through the intersection point. More precisely, we should describe such surfaces as single two-valued surfaces, the two values of the potential energy for each molecular geometry being the eigenvalues of a 2×2 matrix, such as equation (2.3). In more complicated cases we may have n-valued surfaces given by the eigenvalues of an $n \times n$ matrix. The alternative approach is to represent only part of the surface by a single-valued function, or to represent either the top or bottom sheet of an intersection by functions which are non-analytic (i.e. not differentiable) at the crossing point. These possibilities will be explored more fully in later chapters.

2.3 Piecewise or global fitting

A potential energy function and the variables in which it is expressed should be constructed with a particular dynamical problem in mind and with some knowledge of the configuration space covered by the nuclear motion for that problem. There is no point in striving for an accurate fit over the whole configuration space if the dynamical problem is localized in a relatively small region.

It is inevitable that the larger the domain in which one requires a function to be valid, the more complicated will the function have to be. One strategy for overcoming this conflict is to divide up the configuration space by boundaries and to make a piecewise fitting of the potential, using simple functions within each sub-area. If any dynamical problem crosses the boundaries then the functions must be continuous across these and their first derivatives must also be continuous. Without these conditions there would be no continuity of wave functions in a quantum mechanical calculation or conservation of energy in a classical calculation. However, even after the requirements of continuity have been satisfied, the functions within the boundaries can be much simpler (and therefore more economical to use in computation) than a global function which has to fit the whole configuration space.

The most widely explored piecewise fitting function for potential energy surfaces has been the cubic spline. These are cubic polynomials whose coefficients are chosen to ensure continuity of the function and its first two derivatives at the junction points, which are called nodes. The conditions at a node for cubic functions are sufficient to maintain continuity across the boundaries between nodes in the many-dimensional case. Algorithms have been derived for determining node spacing[12]; very roughly, the node density should be proportional to the curvature of the surface.

Sathyamurthy and Raff[13] made an extensive study of cubic spline fitting to

Table 2.1 Standard deviations (σ) of the H_2 potential and its first derivative, for a cubic spline fit to a Morse curve[13]. The Morse function was fitted in the range $0.5 \leqslant R \leqslant 2.5$ Å.

Nodes	$\sigma(V/\mathrm{eV})$	$\sigma(V'/\mathrm{eV\ Å^{-1}})$
10	3.08×10^{-3}	0.101
12	8.26×10^{-4}	0.020
14	4.75×10^{-4}	0.092
16	1.83×10^{-4}	0.067
20	1.44×10^{-4}	0.057

one-, two- and three-dimensional potential energy surfaces. Table 2.1 shows the standard deviations of both the potential and its first derivative for the cubic spline fit to the Morse curve for H_2 as a function of the number of nodes. After a rapid initial improvement the accuracy increases rather slowly as the number of nodes is increased.

In two dimensions they examined a 10×10 and 15×15 rectangular grid of nodes for a collinear H—Cl—H surface and found standard deviations in the potential of $\sigma(V/\mathrm{eV}) = 5.1 \times 10^{-3}$ and 2.4×10^{-3}, respectively. For a $15 \times 15 \times 15$ grid on the equivalent three-dimensional system they found $\sigma(V/\mathrm{eV}) = 0.027$. However, this accuracy was insufficient to give a good match between classical trajectories on the full analytic surface and those on the spline-fitted surface. The main difficulty appeared to lie in the fact that the spline-fitted surface was rather poor in the asymptotic regions (i.e. near to dissociation) where the node density was low.

Following the experience of these authors, the tendency has been against the use of multidimensional splines, and rather to use one- or two-dimensional splines in conjunction with other simple functions (e.g. Morse functions) for the extra dimensions[14]. The method is still being explored. Our view at this stage is that there does not appear to be any advantage in piecewise fitting over global fitting unless one is seeking the highest accuracy at many points on the surface. For systems with more than three atoms the use of cubic splines appears to be particularly difficult.

2.4 The Wigner–Witmer rules

Wigner and Witmer in 1928 derived rules for determining the type of diatomic molecular state that can result from given states of the separate atoms[15]. It is relatively easy to extend these rules to relate the states of a polyatomic molecule to the states of its fragments.

First we deal with the spin multiplicity. If two fragments have spin quantum numbers S_1 and S_2, the spin of the combined system, S, must satisfy the rule for quantum mechanical addition of angular momentum, namely

$$S = (S_1 + S_2), (S_1 + S_2 - 1), (S_1 + S_2 - 2), \ldots, |S_1 - S_2| \qquad 2.6$$

This rule is useful for determining the asymptotic limits of a potential energy surface. For example, the ground state of HCN is a singlet ($S=0$) and this can dissociate to the ground state of $H(^2S)$ and $CN(^2\Sigma^+)$ but it cannot give the ground state of $CH(^2\Pi)$ and $N(^4S)$ because a doublet ($S=\frac{1}{2}$) and a quartet ($S=\frac{3}{2}$) state when combined can, by equation (2.6), give only triplet ($S=1$) and quintet ($S=2$) states. The dissociation of the ground state of HCN into CH and N must give one of these fragments in an excited state.

The spatial symmetry of the electronic state of a diatomic molecule is described by a quantum number Λ, which is the modulus of the orbital angular momentum along the internuclear axis. States are labelled according to this quantum number by the scheme

$$\Lambda = 0, 1, 2, 3, \ldots$$

$$\text{Symbol} = \Sigma, \Pi, \Delta, \Phi, \ldots$$

Σ states are non-degenerate; all others are double degenerate, corresponding to the two senses in which a non-zero orbital angular momentum can be directed along the internuclear axis. Two atoms in S states can only give diatomic Σ states. One atom in an S state and one in a P state can give diatomic Σ and Π states. This is because all the angular momentum arises from the atomic P state which has angular momentum components $M_L = 1, 0, -1$ along the internuclear axis, and $\Lambda = |M_L|$.

Σ states are divided into two classes, Σ^+ and Σ^-, according to whether the state is symmetric ($+$) to reflection in a plane containing the nuclei or antisymmetric ($-$). A single unpaired electron in a σ orbital gives rise to a Σ^+ state; Σ^- states arise, for example, from two unpaired electrons whose angular momenta about the internuclear axis cancel. Herzberg[16] gave the correlation rules for all common pairs of combining atomic states.

For polyatomic molecules the spatial symmetry rules for dissociation are generally less restrictive than for diatomics because their symmetry is lower. The only spatial symmetry operation that exists for all configurations of a triatomic molecule is reflection in the plane of the nuclei. Thus, at a general point on a triatomic surface the molecule belongs to the C_s symmetry group which has irreducible representations A' (symmetric to reflection) and A'' (antisymmetric to reflection). Diatomic Σ^+ states transform as A' in the C_s group, whereas Σ^- transform as A''. All others give one A' and one A'' state. To find the overall designation of the triatomic state in C_s the diatomic symmetry must be combined as a direct product with the atomic symmetry: an S_g state is A', S_u is A'' and all others have at least one A' and one A'' component. For example, $H(^2S_g)$ and $CN(^2\Sigma^+)$ will give only A' states of HCN; $CH(^2\Pi)$ and $N(^4S_u)$ would give both A' and A'' states.

For points on the potential surface where the molecule has higher symmetry than C_s there will be more stringent symmetry rules. For example, the ground state of H_2O is A_1 in C_{2v}, which on distortion becomes A' in C_s. If the molecule is taken to a linear configuration ($C_{\infty v}$ or $D_{\infty h}$ groups) the state becomes Σ^+. If we now consider dissociation to H and OH we have a different situation for linear configurations to non-linear. The ground state of OH is $^2\Pi$ and for non-linear configurations OH($^2\Pi$) plus H(^{2}S) will give one A' and one A'' state. Thus $H_2O(^1A')$ can dissociate to ground-state products. However, in linear configurations OH($^2\Pi$) plus H(^{2}S) will give a Π state, and this is not the symmetry of the ground state of H_2O. Thus if H_2O is first made linear it cannot then dissociate to give the ground state of OH. This example will be examined in greater detail in a later chapter.

References

1. J. N. Murrell and K. Laidler, *Trans. Faraday Soc.*, **64**, 371 (1968).
2. J. von Neumann and E. P. Wigner, *Z. Phys.*, **30**, 467 (1929).
3. C. A. Mead, *J. Chem. Phys.*, **70**, 2276 (1979).
4. H. Rosenthal, *Phys. Rev. A*, **4**, 1030 (1971).
5. E. Teller, *J. Phys. Chem.*, **41**, 109 (1937).
6. H. C. Longuet-Higgins, *Proc. Roy. Soc. London, Ser. A*, **344**, 147 (1975).
7. G. Herzberg and H. C. Longuet-Higgins, *Discuss. Faraday Soc.*, **35**, 77 (1963).
8. I. B. Bersuker and V. Z. Polinger, *Adv. Quantum. Chem.*, **15**, 85 (1982).
9. H. C. Longuet-Higgins, U. Opik, M. H. L. Pryce and R. A. Sack, *Proc. Roy. Soc. London, Ser. A*, **244**, 1 (1958).
10. A. J. C. Varandas, J. Tennyson and J. N. Murrell, *Chem. Phys. Lett.*, **61**, 431 (1979).
11. W. Lichten, *Phys. Rev.*, **131**, 229 (1963); B. C. Garrett and D. G. Truhlar, *Theor. Chem.*, **6A**, 215 (1981).
12. A. R. Curtis, in *Numerical Approximations to Functions and Data*, Ed. J. G. Hayes, Athlone Press, London, 1970, Ch. 4.
13. N. Sathyamurthy and L. M. Raff, *J. Chem. Phys.*, **63**, 464 (1975).
14. S. K. Gray and J. S. Wright, *J. Chem. Phys.*, **66**, 2867 (1977).
15. E. P. Wigner and E. E. Witmer, *Z. Phys.*, **51**, 859 (1928).
16. G. Herzberg, *Spectra of Diatomic Molecules*, Van Nostrand, Princeton, 1950.

Chapter 3

Quantum mechanical calculations and general strategy

3.1 *Ab initio* calculations

We can obtain information about potential energy surfaces either from experimental techniques such as spectroscopy, gas kinetics and molecular beam scattering, or by calculating approximate solutions of the electronic Schrödinger equation. Experimental techniques usually provide information about special local areas of the surface whereas calculations are, in principle, capable of providing information on the whole surface. However, except for the simplest systems, calculations cannot provide data with the accuracy achieved by the best experiment, so that one should look to calculations primarily to provide information on those parts of the surface which are not easily probed by experiment.

The history of quantum mechanical calculations on potential surfaces goes back to the early days of quantum mechanics with studies of the H_3 surface in the 1930s. A simple theoretical model is sufficient to show that the isotope exchange reaction

$$D + H_2 \rightarrow H + HD$$

goes through a collinear transition state in which the D–H and H–H distances are equal, a fact which is not easily deduced from experiment. There are now very accurate calculations on this surface and we shall discuss some of them fully in Chapter 8.

Until the mid 1960s, calculations on potential energy surfaces were based largely on model Hamiltonians or considerable approximations were made in solving the Schrödinger equation for the exact Hamiltonian. With the advent of large digital computers the numerical techniques available for handling the exact Hamiltonian have improved dramatically, and there is now a wealth of such *ab initio* data in the literature. It would be out of place in this book to give a detailed account of these techniques[1–5], but we shall make references to particular methods in later chapters and some basic terminology is necessary.

Almost all calculations employ molecular orbital (MO) theory at the self-consistent-field (SCF) level or go beyond that to include configuration interaction (CI). An orbital is the wave function of one electron in an atom or molecule. It is determined by the field of the nuclei and the field arising from the average distribution of all the electrons. In the SCF method there is consistency between the wave functions and the field deduced from these wave functions. A configuration is an enumeration of the orbital occupations. For example, the ground-state configuration of the carbon atom is $1s^2 2s^2 2p^2$. An SCF wave function is associated with a single configuration. If the configuration is that of a closed shell (all orbitals are either completely filled by electrons or are empty) then the SCF wave function is an antisymmetrized product of orbital and electron spin functions (α or β). This antisymmetrized product can be represented in determinantal form known as a Slater determinant. For open shells it may be necessary to have more than one such determinant in the wave function to obtain the correct total spin symmetry.

The best SCF wave function for a specified configuration is called the Hartree–Fock function. By the variation theorem such a function minimizes the energy of the state for this restricted class of function. SCF wave functions generally give reasonably good relative energies between states of the same molecule or states of different molecules which have the same number of electron pairs (i.e. orbitals containing two electrons). However, they give poor values for the relative energies of states with different numbers of electron pairs, and this usually means that dissociation energies for covalent bond breaking are poor. To obtain satisfactory dissociation energies one must, in general, go beyond the SCF level. This can be done by using multi-configuration SCF functions (these are functions that are variationally optimized for a specified set, generally small, of configurations), or by CI, which also implies a variational optimization for a set of configurations, generally large. For MCSCF wave functions the orbitals are optimized for the whole wave function. In the CI method the orbitals are usually only optimized for the SCF function which is the dominant configuration. Most of the accurate calculations which aim for a wide coverage of the potential surface have been based on SCF wave functions with CI.

Molecular orbitals (ψ_k) are usually obtained as a linear combination of specified functions ϕ_μ, which is called the basis

$$\psi_k = \sum_\mu c_{k\mu}\phi_\mu \qquad 3.1$$

The functions ϕ_μ are often associated with atomic orbitals of the component atoms and equation (3.1) is then known as the linear combination of atomic orbitals (LCAO) approximation. If there is one function for each of the atomic orbitals which is occupied in the ground-state configuration of each atom (e.g. a 1s orbital for H, a 1s, 2s and three 2p orbitals for C), this is referred to as a minimal basis. If there are two functions ϕ_μ for each such atomic orbital, then the basis is referred to as double-zeta quality. Functions which do not appear in the ground-state configurations of the atoms (e.g. 2p orbitals for H or 3d orbitals for C) are called polarization functions. A basis which is of double-zeta quality and includes polarization functions is generally sufficient to give a close approximation to the Hartree–Fock limit.

The explicit functions used for an atomic orbital basis are either exponential functions:

$$\phi_{nlm} = r^n \exp(-\zeta r) Y_{lm}(\theta, \phi) \qquad 3.2$$

known as Slater-type orbitals (STOs), or Gaussian functions:

$$\phi_{nlm} = r^n \exp(-\zeta r^2) Y_{lm}(\theta, \phi) \qquad 3.3$$

known as Gaussian-type orbitals (GTOs). In both cases the functions are the product of a radial function (r is the distance of the electron from the nucleus) and an angular function Y_{lm} which is a spherical harmonic. STOs are more accurate representations of exact atomic orbitals but a GTO basis is much more economical to use in the calculations necessary to produce SCF or SCF CI wave functions.

Most current calculations on polyatomic molecules use a GTO basis, but this is sometimes expressed formally as a STO expansion in which each STO is approximated as a small expansion in GTOs. For example, the symbol STO-3G for a basis means that each STO is approximated by a linear combination of three GTOs, the coefficients in the sum being chosen to minimize the difference between the STO and its STO-3G approximation. There are many GTO expansions in current use which have been optimized for molecular calculations.

Many computer programs have been written for carrying out *ab initio* SCF calculations, and most can be treated as 'black boxes' and used by scientists who have little detailed knowledge of the underlying theory. MCSCF or SCF CI programs are much less widely available and do generally require expert knowledge for their efficient use. It follows that *ab initio* calculations which give a good representation of a potential energy surface can, in general, only be carried out by those with an expert knowledge of quantum chemistry. However, the position is rapidly changing and 'black box' programs which achieve this objective will no doubt be available shortly.

The main difficulty in determining accurate potential energy surfaces from *ab initio* calculations is that even with elaborate calculations it is difficult to obtain a uniform accuracy over the whole surface. Absolute accuracy, as measured by the total binding energy of the electrons to the nuclei, is not required, but relative accuracy, i.e. a constant error in the total energy at all points on the surface, is necessary in order for barrier heights, dissociation energies, etc., to be accurate.

Even if the same basis and the same level of configurational mixing is used for all points on the surface, this does not ensure constant accuracy although, as a general rule, the larger the basis and the larger the extent to which configuration mixing is allowed, the more nearly will constancy be achieved. Unfortunately, one is frequently faced with the problem of matching points on the surface that have been calculated using different bases or different levels of configurational mixing. This is simply because a complete coverage of the surface is a substantial task and until recently quantum chemists have usually confined their attention to limited parts of the surface. For example, there are very many calculations on the ground-state potential of water at, or close to, the equilibrium configuration. There have been several studies on structures of C_{2v} symmetry (both O—H bonds of equal length), but few calculations at other points on the surface.

It is clearly necessary, but unfortunately not easy, to bring 'less accurate' calculations to the same standard as 'more accurate' calculations before an overall fitting to some functional form of the potential can be attempted. In addition, if there are regions of the surface which lack data of any kind then it may be necessary to make a local interpolation of surrounding data into that region in order to provide some restraint on any global function. Without this, the global function may take unphysical character in an unrestrained region.

3.2 The LEPS and related functions

The second theme we wish to explore is the use of functions which are based on a simple quantum mechanical description of the electronic wave function for the state in question. This approach has its roots in the early study of H_3 by London[6] and has developed into the very widely used LEPS (London, Eyring, Polanyi, Sato) function for interacting doublet-state atoms. In the Heitler–London treatment of the hydrogen molecule, the interaction of two ^2S atoms gives rise to a singlet ground state and triplet excited state with energies (relative to separated atoms)

$$E_S = (Q + A)(1 + S^2)^{-1}$$
$$E_T = (Q - A)(1 - S^2)^{-1} \qquad 3.4$$

where Q is the coulomb integral:

$$Q = \iint 1s_a^2(1) H' 1s_b^2(2)\, dv_1 dv_2 \qquad 3.5$$

A is the exchange integral:

$$A = \iint 1s_a(1) 1s_b(1) H' 1s_a(2) 1s_b(2)\, dv_1 dv_2 \qquad 3.6$$

and S is the overlap integral:

$$S = \int 1s_a(1) 1s_b(1)\, dv_1 \qquad 3.7$$

H' is the interaction Hamiltonian, which is the difference between the total Hamiltonian and the Hamiltonian for the two separated atoms:

$$H' = -r_{b1}^{-1} - r_{a2}^{-1} + r_{12}^{-1} + r_{ab}^{-1} \qquad 3.8$$

and $1s_a$ and $1s_b$ are 1s orbitals of the two atoms.

The coulomb and exchange integrals are both negative quantities for values of the internuclear distance r_{ab} greater than (roughly) the equilibrium distance of the ground state of H_2, and $|A| \gg |Q|$. It follows that E_S is negative and E_T is positive; at short internuclear distances both energies become positive.

From the wave functions of three ^2S atoms one can construct one quartet and two doublet states. For H_3 the wave functions of the two doublet states are, in the Heitler–London formulation

$$\Psi_1 = \sqrt{\tfrac{1}{2}}\mathscr{A}\{1s_a(1)1s_b(2)1s_c(3)[\alpha(1)\alpha(2)\beta(3) - \alpha(1)\beta(2)\alpha(3)]\}$$

$$\Psi_2 = \sqrt{\tfrac{1}{6}}\mathscr{A}\{1s_a(1)1s_b(2)1s_c(3)[2\beta(1)\alpha(2)\alpha(3) - \alpha(1)\alpha(2)\beta(3) - \alpha(1)\beta(2)\alpha(3)]\} \qquad 3.9$$

where $\mathscr{A}$ is the antisymmetrizing operator. If overlap integrals are neglected the 2×2 matrix of the interaction Hamiltonian has the following elements:

$$H'_{11} = Q_{ab} + Q_{bc} + Q_{ca} + A_{bc} - \tfrac{1}{2}A_{ab} - \tfrac{1}{2}A_{ac}$$

$$H'_{22} = Q_{ab} + Q_{bc} + Q_{ca} - A_{bc} + \tfrac{1}{2}A_{ab} + \tfrac{1}{2}A_{ac}$$

$$H'_{12} = \sqrt{\tfrac{3}{4}}(A_{ab} - A_{ac}) \qquad 3.10$$

The eigenvalues of this matrix are

$$E_{\pm} = Q_{ab} + Q_{bc} + Q_{ca} \pm \sqrt{\tfrac{1}{2}}\,[(A_{ab} - A_{bc})^2 + (A_{bc} - A_{ca})^2 + (A_{ca} - A_{ab})^2]^{1/2} \qquad 3.11$$

which, having the general form

$$E_{\pm} = X \pm Y^{1/2} \qquad 3.12$$

is a two-valued function with a degeneracy when $Y=0$; it can be seen that this occurs for D_{3h} structures where $A_{ab} = A_{bc} = A_{ca}$. Comparison with equation (3.4) ($S=0$) shows that expression (3.11) is not equal to the sum of three H_2 pair potentials.

If overlap is not neglected the expressions for the matrix elements (3.10) become extremely complicated even for this simple system of three hydrogen atoms. The complete expressions have been given by Slater[7] and by Eyring and Lin[8] and have been used semi-empirically by Porter and Kaplus[9]. However, Sato[10] proposed that the effects of overlap could be introduced simply by multiplying expression (3.11) by a factor $(1 \pm k)^{-1}$.

By treating k as an empirical parameter, independent of the internuclear distances, Sato was able to obtain a semi-empirical potential energy function that has been widely used for dynamical studies. With only minor modifications this has become the LEPS function.* It must be emphasized that Sato did not give a mathematical justification of his procedure, and Eyring and Lin[8], who examined the problem in detail, concluded that no rigorous justification can be given.

The standard procedure which is followed to obtain a LEPS surface is to start with expressions for the potential of the diatomic singlet and triplet states. The singlet potential is usually taken as a Morse function (1.2) and the triplet as an

* Eyring and Polanyi's contribution was to give the first semi-empirical treatment of the London expression (3.11)[11].

anti-Morse function [as (1.2) but with both exponentials taken with positive signs]. Other more accurate functions could equally well be used. The essential point of the treatment is that Q_{ab} and A_{ab}, which appear in equation (3.11), are obtained by substituting the diatomic potentials into equation (3.4), and replacing S^2 by the Sato parameter k. The same value of k is then used in the multiplication factor $(1 \pm k)^{-1}$ which is applied to the triatomic potential (3.11), and its value is chosen empirically to reproduce some important feature of the surface such as the height of a saddle point.

Although expression (3.11) has been devised for the doublet surfaces of three interacting ^2S atoms [this is based on the assumption that only the 'covalent' structures Ψ_1 and Ψ_2 (3.9) are important], it has also been used for the interaction of other atoms with doublet-spin ground states such as the halogens. However, these have ^2P ground states, and there are therefore more than two doublet surfaces which dissociate to ground-state atoms.

The LEPS function derived by the method described above has several good features, not least that it dissociates to the correct diatomic potentials and has the correct topographical features for two intersecting doublet surfaces. However, it has the disadvantage that it contains only one parameter (k) which can be adjusted to the properties of the triatomic surface. Further flexibility has been built into the method by several authors, and some examples of this will be discussed further in Section 6.2 and Chapter 8.

Although the valence-bond expression for the doublet state energies (3.11) gives both sheets of the two-valued surface, for dynamics one is usually only interested in the lower energy sheet E_-. For this reason there is no need to have an accurate representation of the diatomic triplet potential because this is not a product of dissociation of the lower surface. The expression for the diatomic triplet potential can therefore be modified to improve the properties of the triatomic surface and one can justify this as an effective contribution of other valence bond configurations, e.g. of ionic states. A modified LEPS function for H_3 derived in this way will be described in Section 8.3.

3.3 The diatomics in molecules (DIM) method

The DIM method is a semi-empirical valence bond method of calculating the potential energies of small polyatomic molecules. It goes further than the LEPS method described in the last section because it is applicable to surfaces formed from atoms of any spin or spatial symmetry. In its simplest form, for three ^2S atoms, it gives a surface equivalent to that obtained from the London equation.

The method was first proposed by Ellison[12] as an extension of Moffitt's 'atoms-in-molecules' method[13] and it has some similarities with other theories of weak interatomic or intermolecular interaction[14]. The common theme in all of these is that the electronic states of a molecule have some relationship to the states of its component atoms or sub-molecular fragments and the relationship is close in the region of weak interaction. The DIM method has been reformulated by several authors and we give only a brief summary here. The reader is referred to these specialist articles for details of the method[15,16].

The DIM wave functions are constructed from products of atomic basis functions (spin orbitals), ϕ_μ, of the atoms which are in the molecule.

$$\psi_i = \phi_\mu^{(A)} \phi_\nu^{(B)} \cdots \phi_\sigma^{(N)} \tag{3.13}$$

These are then antisymmetrized by an operator $\mathscr{A}$ which produces all permutations of electrons amongst the basis functions with the appropriate sign:

$$\Psi_i = \mathscr{A} \psi_i \tag{3.14}$$

These functions are closely connected to valence bond functions, which are either the individual functions Ψ_i or linear combinations of degenerate Ψ_i which are necessary to give appropriate spin and orbital angular momentum eigenfunctions. The selection of the DIM basis is made on the same principles as one selects valence bond functions; each function is associated with a classical valence structure (covalent, ionic, etc.) and chemical principles are used to select the structures of lowest energy. As a minimum requirement for the DIM basis one needs those functions which, when the molecule dissociates, represent the wave functions of the open channels of interest.

A typical Hamiltonian matrix element in the basis Ψ is

$$H_{ij} = \langle \Psi_i | H | \Psi_j \rangle \tag{3.15}$$

but, as H is symmetric in the electrons, H and $\mathscr{A}$ commute and hence equation (3.15) can be rewritten as

$$H_{ij} = \langle \Psi_i | \mathscr{A} H \psi_j \rangle \tag{3.16}$$

As ψ_j implies a particular allocation of electrons to atoms it is useful to separate H into terms which are consistent with this allocation. This can be done in several ways; for example, one can separate the Hamiltonians for the separate atoms from the interaction terms, as in the atoms-in-molecules method[13]. The particular partition used in the DIM method is

$$H = \sum_{A} \sum_{B>A} H^{(AB)} - (N-2) \sum_{A} H^{(A)} \tag{3.17}$$

where N is the number of atoms. $H^{(AB)}$ is the Hamiltonian for the AB diatomic molecule and, as this also contains the atomic terms $H^{(A)}$ and $H^{(B)}$, their subtraction with a factor $(N-2)$ is required to prevent overcounting.

With the partition (3.17), the matrix element (3.16) can be separated into a sum of terms. Some of these involve the atomic Hamiltonians

$$H_{ij}^{(A)} = \langle \Psi_i | \mathscr{A} H^{(A)} | \psi_j \rangle \tag{3.18}$$

and some the diatomic Hamiltonians

$$H_{ij}^{(AB)} = \langle \Psi_i | \mathscr{A} H^{(AB)} | \psi_j \rangle \tag{3.19}$$

Because $H^{(A)}$ and $H^{(AB)}$ are not symmetric in all the electrons, they do not commute with $\mathscr{A}$, hence the components of the Hamiltonian matrix (3.18) and (3.19) are not Hermitian. Although the whole matrix, if calculated exactly, is necessarily Hermitian, if approximations are made to the components of the matrix elements then its Hermitian property may be lost.

The first step in evaluating equation (3.18) is to note that $H^{(A)}$ operates only on the atomic basis functions of A which are in ψ_j; likewise in equation (3.19), $H^{(AB)}$ only operates on the products $\phi_\mu^{(A)}\phi_\nu^{(B)}$ which are in ψ_j. As a general result we can write

$$H^{(A)}\phi_\mu^{(A)} = \sum_\lambda d_{\mu\lambda}\phi_\lambda^{(A)} \qquad 3.20$$

if $\phi_\lambda^{(A)}$ is a complete set of states for the operator $H^{(A)}$. An important approximation in the DIM method is that equation (3.20) is exact even when the $\phi_\lambda^{(A)}$ are restricted to the finite number of basis functions which are being used to construct the DIM wave functions. In the extreme case that only one function is being used for atom A then this approximation is equivalent to assuming that $\phi_\mu^{(A)}$ is an eigenfunction of $H^{(A)}$, and $d_{\mu\mu}$ is its eigenvalue. In any case, a variational calculation on $H^{(A)}$ with the finite basis will give eigenvalues and eigenvectors of the $H^{(A)}$ matrix, from which, by matrix inversion, the coefficients that occur in equation (3.20) can be determined.

In a similar manner, the effect of $H^{(AB)}$ on a product $\phi_\mu^{(A)}\phi_\nu^{(B)}$ can be written as a sum of terms in the products of $\phi_\lambda^{(A)}\phi_\sigma^{(B)}$, and if this is assumed to be exact for the finite set then the coefficient of this expansion can be obtained by carrying out a variational calculation on the diatomic AB. The equations needed to carry out these manipulations are given in detail in appropriate reviews[16].

The empirical approach commonly adopted in the DIM method is to take the eigenvalues of $H^{(A)}$ as the experimental (spectroscopic) energies of A and the eigenvalues of AB as the potential energy curves of the AB molecule; these can be determined by spectroscopy or by accurate *ab initio* calculations. The eigenvectors of these operators are either unit matrices [for $H^{(A)}$] or the coefficients of simple valence bond functions [for $H^{(AB)}$]. With this procedure the DIM potential will have the correct atomic and diatomic dissociation limits and, if sufficient basis functions have been used to construct the DIM Hamiltonian matrix, the potential should be approximately correct in the interaction region. If a small basis is used then the interaction potential may not be reliable because the basis will not be sufficiently flexible to allow for distortions and reorganization of the electron wave functions that accompany bond formation.

It must be pointed out that if more than two basis functions are used to construct the DIM matrix, then its eigenvalues (the DIM potentials) cannot generally be expressed in analytical form. Thus, as with any other method for calculating points on a potential energy surface, it will generally be necessary to find an analytical fit to the eigenvalues if one wishes to use the potential in a dynamic calculation. It is also relevant to note that for more than three atoms the DIM method becomes very complicated and has not found general favour.

There have been several attempts to improve the DIM method by giving it additional flexibility when a small number of basis functions are used. These methods either involve modification of the DIM matrix elements, as in Wu's generalized DIM theory[17] (which returns to Moffitt's separation of the Hamiltonian into atomic and interaction terms), or the addition of a three-body correction as recently proposed by Last and Baer[18].

3.4 The many-body expansion

We have already commented on the fact that the LEPS function for a triatomic potential energy surface is not the sum of atom–atom pair interactions. This is not surprising because it is a characteristic of chemical bonding that the properties of an atom depend on the other atoms to which it is attached. In fact, apart from inert gas clusters there has been very little inclination to discuss the potential energies of molecular species by making an expansion in the energies of sub-clusters of atoms. However, we shall show that there are advantages to this approach even when the expansion does not show rapid convergence.

We write the many-body expansion for a single-valued potential energy surface as follows[19–21]:

$$V_{\mathrm{ABC}\cdots N}=\sum V_{\mathrm{A}}^{(1)}+\sum V_{\mathrm{AB}}^{(2)}(R_{\mathrm{AB}})+\sum V_{\mathrm{ABC}}^{(3)}(R_{\mathrm{AB}},R_{\mathrm{BC}},R_{\mathrm{CA}})+\cdots V_{\mathrm{ABC}\cdots N}^{(n)}(R_i) \tag{3.21}$$

$V_{\mathrm{A}}^{(1)}$ is the energy of atom A in the state which is produced by adiabatically removing this atom from the cluster. We are here assuming that there is only one state produced by this dissociation and will later discuss more complicated examples where this is not true. $\sum V_{\mathrm{A}}^{(1)}$ is the sum of all one-body terms. If the energy zero is chosen with all atoms in their ground states, then $V_{\mathrm{A}}^{(1)}$ is only non-zero if, on dissociation, atom A is left in an excited state.

$V_{\mathrm{AB}}^{(2)}(R_{\mathrm{AB}})$ is a two-body term which is a function of the separation of the two atoms and which tends asymptotically to zero as R_{AB} approaches infinity; $\sum$ indicates the sum of all such terms.

$V_{\mathrm{ABC}}^{(3)}$ is a three-body energy which depends on the dimensions of the ABC triangle. We have indicated its functional dependence on the three internuclear distances but any other set of independent coordinates could be chosen. A three-body term must become zero if any of the three atoms is removed to infinity, and this condition is likely to be most easily imposed by using the internuclear distances as variables.

The last term in the expansion is the n-body term for an n-atom cluster. This term will become zero if any one of the atoms is removed to infinity. $V^{(n)}$ is, therefore, like the complete potential, a function of the $3n-6$ internal coordinates of the system, and forming the many-body expansion leads to no simplification of the problem unless at least one of the following two situations prevails: (a) the series converges rapidly and some of the high-order terms can be neglected; (b) there exist simple functional forms for the higher terms in the expansion.

There are two factors pertinent to the convergence of the expansion. Firstly, there is the magnitude of a term $V^{(r)}$ compared with $V^{(r+1)}$, and secondly there is the magnitude of the total of $V^{(r)}$ terms compared with the total of $V^{(r+1)}$ terms. Regarding the first of these, it is highly probable that for most configurations of relevance to problems of chemical dynamics (for bound states and for collisions of atoms and molecules with less than 10 eV of translational energy) the convergence is rapid after $V^{(3)}$. The arguments for this statement rest on the few tetra-atomic molecules whose potential surfaces have been studied and on the quantum mechanical interpretation of an n-body term. The situation regarding the second factor is more difficult to judge. There are $n!/r!\,(n-r)!$ r-body terms in an n-atom system and although we might be tempted to ignore, say, four-body terms, there are of the order $n^4/24$ of them when n is large.

It is fair to say that except for inert gas clusters there is at present no evidence to say where, or even if, a many-body expansion can be terminated. For small polyatomic molecules (those with three or four atoms), it is necessary to retain all terms if chemical accuracy is desired. We must therefore look to the feature (b) above to justify using the many-body expansion for such systems.

Although the dimensionality of an r-body term increases with increasing r, there are reasons to believe that they may become simpler functions of the internal coordinates. Firstly, as was pointed out by Vance and Gallup[22], the total potential has singularities at $R_i=0$ arising from the nuclear–nuclear repulsion, but these singularities are already contained in the two-body terms. The remaining evidence is circumstantial but not overwhelming. In the few tetra-atomic surfaces that have been examined in detail[19,23,24] the main features of the surface appear to be contained in the two- and three-body terms, and the four-body term can be regarded as fine tuning to give chemical accuracy. In all such cases studied the four-body terms have been given relatively simple forms.

There is also evidence from *ab initio* calculations of the equilibrium geometries and energies of X_2, X_3 and X_4 systems that the two- and three-body terms are predominantly structure determining in elemental clusters[25]. For example, Be_2 and Be_3 are very weakly bonded but Be_4 has a fairly stable tetrahedral structure. At first sight this might appear to be due to a strongly stabilizing four-body term, but it is in fact due to a strongly stabilizing three-body term. In Be_3, whose structure is an equilateral triangle, the three repulsive two-body energies roughly balance the single attractive three-body energy. On going from Be_3 to Be_4 the two-body terms are doubled in number but the three-body terms are quadrupled and the attractive terms therefore predominate. The four-body term in Be_4 is in fact slightly repulsive[26]. There are recent calculations on X_3 and X_4 systems for the elements N, O, P and S which confirm the general principle that four-body terms may contribute significantly to the total energy but are not structure determining[27–29]. The difference between O and S clusters, for example, resides in the fact that the three-body terms are repulsive for O but attractive, or much less repulsive, for S[25,29].

Suppose finally that the many-body expansion is not rapidly convergent, nor are the terms functionally simple; are there any advantages left? We think there are. Firstly, the potential written in this form is guaranteed to satisfy all dissociation limits. For example, if we consider the reaction

$$ABCD \rightarrow AB + CD$$

then for the products, the only terms in the many-body expansion which will be non-zero are, apart from the one-body terms, $V_{AB}^{(2)}(R_{AB})$ and $V_{CD}^{(2)}(R_{CD})$, which is as required.

The second advantage is that the many-body expansion suggests a strategy for building up a polyatomic potential by studying the potentials of all the fragments. Thus, the two-body terms can be deduced from diatomic potentials and the three-body terms are given by the difference between triatomic potentials and the sum of the one- and two-body components, i.e.

$$V_{ABC}^{(3)} = V_{ABC} - \sum V_A^{(1)} - \sum V_{BC}^{(2)} \qquad 3.22$$

Thus, we can make a first estimate of the potential energy surface of H_2CO, say, by knowing the potentials for the relevant states of the diatomic species H_2, CH, OH and CO, and having obtained three-body terms for the relevant states of H_2O, H_2C and HCO.

Although this seems to be a great deal of work to obtain one tetra-atomic potential, it must be remembered that all of these terms are needed to give a full account of all the dissociation limits. Moreover, these terms, once deduced, will be used not just for H_2CO but also for many other tetra-atomic (or larger) species containing the same elements. One can therefore envisage a data bank of two- and three-body terms which will, in various combinations, provide the first approximation to any tetra-atomic potential. It is with this strategy in mind that we discuss specific examples of tri- and tetra-atomic species in the remaining chapters of this book.

3.5 Functional form of many-body terms

In Section 1.2 we described some of the most common functions that have been used to represent diatomic potentials. Weighing in the balance simplicity and accuracy (achieved through flexibility), our view is that the Extended Rydberg function (1.19) has a good claim to being the most satisfactory function to represent valence (i.e. intermediate and short-range) interactions, and if it is supplemented by an R^{-n} expansion it can encompass the long-range interaction region. From this starting point it is natural to think of representing the many-body terms in a potential by a function which is similar but of greater dimensionality.

The requirement that a many-body term tends asymptotically to zero if one atom in the cluster is removed to infinity is satisfied by writing the term as a product of a polynomial P and some function T, which we call the range function, which tends to zero as an exponentially decaying function of the internuclear distances:

$$V^{(n)} = P(R_i)T(R_i) \qquad 3.23$$

The Extended Rydberg function for $V^{(2)}$ is a one-dimensional function of this form.

The obvious choice for the range function is

$$T = \exp\left(-\sum_{i=1}^{3n-6} \gamma_i \mathrm{R}_i\right) \qquad 3.24$$

where R_i are the set of internuclear distances which define the geometry of the n-atom cluster. However, this function was examined for the H_2O surface[30] and was found to give too rapid a change for the three-body term in some of the physically significant regions of the surface.

For diatomic potentials one is not generally concerned with interatomic distances that are much shorter than equilibrium distances, because in such regions the potential is very repulsive. For polyatomic molecules this is not the case. The equilibrium H–H distance in the water molecule, for example, is 1.5 Å, yet distances of 0.74 Å, the equilibrium distance in H_2, are relevant to physical processes that can occur on the H_2O surface. For this reason the rapid increase which occurs in an exponential function at short distances is a disadvantage. This is supported by our earlier comment that n-body terms ($n > 2$) are less rapidly varying functions than two-body terms because the latter contain the nuclear–nuclear repulsion which is singular at $R_i = 0$.

A function that does not have the above disadvantage of an exponential, yet still behaves as an exponential at large values of R_i, is based on the hyperbolic tangent

$$1 - \tanh x = 1 - \frac{e^x - e^{-x}}{e^x + e^{-x}} = \frac{2e^{-2x}}{1 + e^{-2x}} \qquad 3.25$$

At large positive values of x this function approaches zero as $2e^{-2x}$. As many-dimensional alternatives to equation (3.24) we therefore have either

$$T = \prod_{i=1}^{3n-6} (1 - \tanh \gamma_i R_i/2) = \frac{2^{3n-6} e^{-\Sigma \gamma_i R_i}}{\Pi\,(1 + e^{-\gamma_i R_i})} \qquad 3.26$$

or

$$T = \left(1 - \tanh\left[\tfrac{1}{2} \sum_{i=1}^{3n-6} \gamma_i R_i\right]\right) = \frac{2e^{-\Sigma \gamma_i R_i}}{(1 + e^{-\Sigma \gamma_i R_i})} \qquad 3.27$$

It can be seen that both behave similarly when any R_i approaches infinity, essentially as $\exp\left[-\sum_i \gamma_i R_i\right]$ apart from a multiplying factor. For complete generality one can use a product such as equation (3.26) but with the R_i replaced by $\rho_i = R_i - R_i^0$ which are displacements from some reference structure (R_i^0), or by variables S_i which are orthogonal combinations of ρ_i. For systems with permutation symmetry the S_i variables are defined, in whole or in part, by symmetry, and in this case γ_i will be zero for any S_i which is not totally symmetric to permutation of identical atoms; this eliminates any dependence of non-totally symmetric S_i from the range function.

The problem of determining the functional form of an n-body term therefore falls into three parts. Firstly, the variables S_i can be determined, either by symmetry (Section 4.2) or in terms of turning points on the surface (Section 4.3).

Secondly, the parameters γ_i have to be determined. This involves a non-linear optimization of the surface to some required properties. Thirdly, the coefficients of the generalized polynomial have to be determined. This involves both a selection of terms to be included in the polynomial (following the strategy to be outlined in Section 4.3 and discussed in detail later) and a linear optimization of the coefficients of these terms. We would not pretend that this procedure is simple as it is overall a non-linear variational problem. However, we shall show that it can be completed successfully and with modern computational technology is not too arduous a task.

References

1. H. F. Schaefer, *The Electronic Structure of Atoms and Molecules*, Addison-Wesley, Reading, MA, 1972.
2. L. C. Snyder and H. Basch, *Molecular Wave Functions and Properties*, Wiley, New York, 1972.
3. D. B. Cook, *Ab Initio Valence Calculations in Chemistry*, Butterworths, London, 1974.
4. R. F. W. Bader and R. A. Gangi, *Ab Initio Calculation of Potential Energy Surfaces*, Chemical Society Specialist Periodical Report, *Theoretical Chemistry*, Vol. 2, Chemical Society, London, 1975, p. 1.
5. H. F. Schaefer (Ed.), *Modern Theoretical Chemistry*, Vol. 3, *Methods of Electronic Structure Theory*; Vol. 4, *Applications of Electronic Structure Theory*, Plenum Press, New York, 1977.
6. F. London, *Z. Electrochem.*, **35**, 552 (1929).
7. J. C. Slater, *Phys. Rev.*, **38**, 1109 (1931).
8. H. Eyring and S. H. Lin, in *Physical Chemistry, An Advanced Treatise*, Vol. 6A, Eds. H. Eyring, D. Henderson and W. Jost, Academic Press, New York, 1974.
9. R. N. Porter and M. Karplus, *J. Chem. Phys.*, **40**, 1105 (1964).
10. S. Sato, *J. Chem. Phys.*, **23**, 592, 2465 (1955).
11. H. Eyring and M. Polanyi, *Z. Phys. Chem., Abt. B*, **12**, 279 (1931).
12. F. O. Ellison, *J. Am. Chem. Soc.*, **85**, 3540 (1963).
13. W. Moffitt, *Proc. Roy. Soc. London, Ser. A*, **210**, 245 (1951).
14. J. N. Murrell, 'Intermolecular Energies in the Region of Small Orbital Overlap', Ch. 7 in *Orbital Theories of Molecules and Solids*, Ed. N. H. March, Clarendon Press, Oxford, 1974.
15. P. J. Kuntz, 'Semi-Empirical Atom–Molecule Potentials for Collision Theory', in *Atom–Molecule Collision Theory*, Ed. R. B. Bernstein, Plenum Press, New York, 1979.
16. J. C. Tully, *Diatomics in Molecules, Semi-Empirical Methods of Electronic Structure Calculation*, Vol. 7, in *Modern Theoretical Chemistry*, Ed. G. A. Segal, Plenum Press, New York, 1977.
17. A. A. Wu, *Mol. Phys.*, **38**, 843 (1979).
18. I. Last and M. Baer, *Chem. Phys. Lett.*, **73**, 514 (1980).
19. A. J. C. Varandas and J. N. Murrell, *Faraday Discuss. Chem. Soc.*, **62**, 92 (1977).
20. J. N. Murrell, *Isr. J. Chem.*, **19**, 283 (1980).
21. S. Carter, I. M. Mills and J. N. Murrell, *Mol. Phys.*, **39**, 203 (1980).
22. R. L. Vance and G. A. Gallup, *J. Chem. Phys.*, **69**, 736 (1978).
23. S. Carter, I. M. Mills and J. N. Murrell, *Mol. Phys.*, **39**, 455 (1980).
24. S. Carter, I. M. Mills and J. N. Murrell, *Mol. Phys.*, **41**, 191 (1980).
25. J. N. Murrell, *Chem. Phys. Lett.*, **55**, 1 (1978).
26. W. Kolos, F. Nieves and O. Novaro, *Chem. Phys. Lett.*, **41**, 431 (1976).

27. J. N. Murrell, O. Novaro, S. Castillo and V. Saunders, *Chem. Phys. Lett.*, **90**, 421 (1982).
28. O. Novaro and S. Castillo, personal communication.
29. W. L. Feng and O. Novaro, personal communication.
30. K. S. Sorbie and J. N. Murrell, *Mol. Phys.*, **29**, 1387 (1975).

Chapter 4

Coordinates

4.1 Choice of coordinates

There is no one set of coordinates for defining the geometry of a polyatomic system which is convenient for all types of dynamical calculation. For example, spectroscopists will generally favour valence coordinates, a mixture of bond lengths and bond angles, whereas for non-reactive scattering it is most convenient to use the distance between the centres of mass and the internal coordinates of the colliding molecules. One must be prepared to change potential functions from one coordinate system to another, which is a straightforward but sometimes tedious procedure.

We first examine the situation for a triatomic system. Figure 4.1 shows three possible coordinate systems. In two of these, (a) and (b), one atom has been placed in a different context to the other two. In (a) it is the central atom in the triatomic molecule, the lines conforming to the normal rules for indicating chemical bonds, and in (b) it is a separate atom placed in relationship to the diatomic molecule AB. R in this case is the distance from C to the centre of mass of AB. Only in (c) do we have a coordinate system which treats the three atoms as equivalent. There are, of course, other coordinates which uniquely define a triangle, two angles and one side for example, but those shown are the most important for dynamical problems.

Coordinates (a) are most favoured for analysing the vibrational states of an ABC triatomic molecule because for small displacements from equilibrium the potential energy can, to a first approximation, be written as a sum of quadratic terms in the displacements as follows:

$$2V = k_1\,\Delta r_1^2 + k_2\,\Delta r_2^2 + k_\alpha\,\Delta\alpha^2 \qquad 4.1$$

In this case, with some further assumptions about the kinetic energy of the particles, the three normal modes of vibration of the molecule can be identified as AB stretch, BC stretch and A$\hat{\text{B}}$C bend. For real potentials there will be other terms in the potential representing bend–stretch interaction, etc., but the dominant theme of vibrational spectroscopic analysis is that equation (4.1) is usually a good starting point.

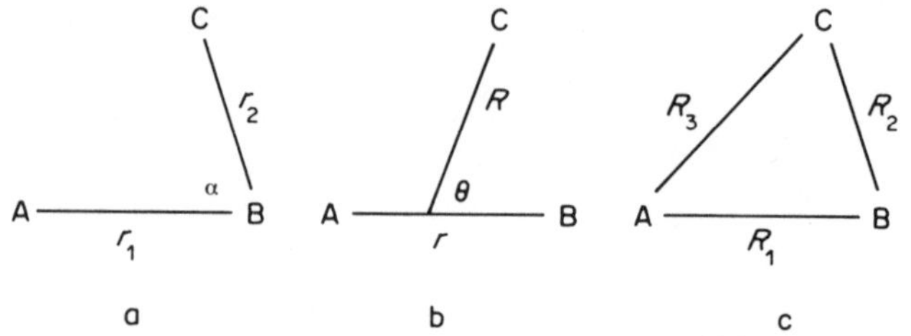

Figure 4.1 Coordinates commonly used for triatomic potentials

In the theory of the non-reactive scattering of an atom and a diatomic molecule, the changes in rotation–vibration states of the diatom in the collision can most simply be calculated by writing the potential in the form

$$V = \Sigma_\lambda F_\lambda(r, R) P_\lambda(\cos\theta) \qquad 4.2$$

where $P_\lambda(\cos\theta)$ is a Legendre polynomial in $\cos\theta$. $F_\lambda(r, R)$ is a function of only the two distance parameters; in the rigid rotor approximation r is fixed at the equilibrium distance of the diatom and F_λ is only a function of R. The reason for using equation (4.2) in scattering calculations is that the coupling potential matrix elements between rotational states J and J' satisfy certain selection rules depending on the values of $|J-J'|$ and λ.

For both spectroscopic and scattering studies it is often sufficient to have a potential which is limited to certain regions of the full configuration space. For example, in non-reactive AB + C scattering it is usually not relevant to know the value of the potential for very large AB distances. For such studies it is most convenient to derive potential functions in one or other of the coordinate systems of Figure 4.1a or b. However, if one wants a potential covering the whole configuration space, which can, in principle, be used for any dynamical study of the ABC system, then it is more sensible to use the even-handed system as in Figure 4.1c. It is with this system, or some orthogonal transformation of it, that we shall be mainly concerned in this book.

Many of the potential functions that will be described later have been derived from spectroscopic data, and an important step in this derivation is the transformation between the coordinate systems (a) and (c) in Figure 4.1. We let $W(r_1, r_2, \alpha)$ be the potential function expressed in valence coordinates (Figure 4.1a), and $V(R_1, R_2, R_3)$ be the potential expressed in the three internuclear distances. The two coordinate systems are related by the cosine rule

$$R_3^2 = r_1^2 + r_2^2 - 2r_1 r_2 \cos\alpha \qquad 4.3$$

and the partial derivatives are related by the expressions given in Appendix 1. It is perhaps worh emphasizing that the two derivatives

$$\left(\frac{\partial W}{\partial r_1}\right)_{r_2\alpha} \quad \text{and} \quad \left(\frac{\partial V}{\partial R_1}\right)_{R_2R_3} \qquad 4.4$$

are not equal; to change r_1 keeping r_2 and α constant requires a completely different displacement of the atoms to changing R_1 keeping R_2 and R_3 constant.

If an analytical expression is known for the potential $V(R_1, R_2, R_3)$ then the terms in the expansion of the scattering potential (4.2) can be obtained by writing

$$\Sigma_\lambda F_\lambda(r, R)P_\lambda(\cos\theta) = V(R_1, R_2, R_3) \tag{4.5}$$

On multiplying both sides of this expression by one of the Legendre polynomials, say, $P_\mu(\cos\theta)$, and making use of their orthogonality, we have

$$F_\mu(r, R) = \tfrac{1}{2}(2\mu+1)\int_0^\pi V(R_1, R_2, R_3)P_\mu(\cos\theta)\sin\theta\, d\theta \tag{4.6}$$

R_2 and R_3 can be replaced by R and θ using the following relationships between the coordinate systems of Figures 4.1b and c:

$$R_2^2 = R^2 - \frac{2m_A}{m_A+m_B} rR\cos\theta + \left(\frac{m_A}{m_A+m_B}\right)^2 r^2 \tag{4.7}$$

$$R_3^2 = R^2 + \frac{2m_B}{m_A+m_B} rR\cos\theta + \left(\frac{m_B}{m_A+m_B}\right)^2 r^2 \tag{4.8}$$

As the scattering coordinates defined in Figure 4.1b distinguish one atom, C, from the other two, A and B, their use in reactive scattering calculations (AB + C $\rightarrow$ A + BC) requires one to make a coordinate transformation between the reactant and product channels. To avoid this complication whilst retaining the advantage of scattering coordinates, namely the separation of the internal variable r from the external variables (R, θ), Marcus[1,2] introduced the concept of natural reaction coordinates. These were based on the curvilinear reaction path which follows the bottom of the potential surface from the reactant to the product valley and for a repulsive surface passes through the saddle point. If C is the curve which describes this path, then the natural reaction coordinates of a point P (p, s) are defined by the length of the normal from P to C, and the distance s along C to this normal (see Figure 4.2).

For a collinear triatom surface A–B–C, the A–B and B–C distances are given relative to their equilibrium values in the diatomic molecules, r^0 by[3]

$$x = r_{AB} - r_{AB}^0$$
$$y = r_{BC} - r_{BC}^0 \tag{4.9}$$

By introducing mass weighted coordinates (r, R) by

$$r = x$$
$$R = \frac{m_A+m_B}{(Mm_Am_B/m_C)^{1/2}}\left(y + \frac{m_A}{m_B+m_C}x\right) \tag{4.10}$$

where

$$M = m_A + m_B + m_C \tag{4.11}$$

one eliminates cross terms in the kinetic energy between $\dot{r}$ and $\dot{R}$. The asymptotic reaction paths of the reactants and products in this new coordinate system are not orthogonal but subtend an angle θ, where

$$\tan \theta = \left(\frac{m_B M}{m_A m_C}\right)^{1/2} \qquad 4.12$$

as illustrated in Figure 4.2.

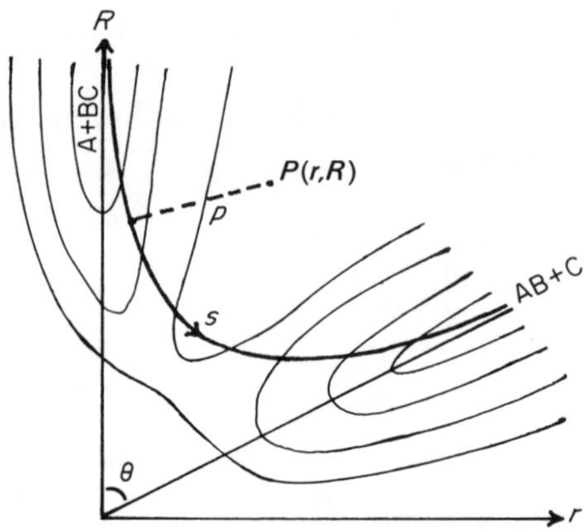

Figure 4.2 Natural reaction coordinates (p, s) for a collinear triatomic system

One disadvantage of the Marcus system of coordinates is that a point on the surface may not have unique coordinates (p, s) because there may be more than one perpendicular p from a point to the curve C. This is particularly true for points far from C and if the curvature of C is high. For this reason Light[3] suggested that C should not necessarily coincide with the reaction path as normally defined but may be any less strongly curved path which has the same asymptotic limits. In particular he recommends the curve

$$R = \frac{\gamma}{r} + r \cot \theta \qquad 4.13$$

where γ determines the curvature of the hyperbolic path between the reactant and product valleys. The distance along the curve C of the perpendicular from the point (r, R) is given by

$$s = s_0 - \frac{\gamma}{r_c} + (1 + \cot^2 \theta)^{1/2} r_c \qquad 4.14$$

where r_c is the value of r on the curve at the foot of the perpendicular.

Natural reaction coordinates have been extensively used in quantum collinear reactive scattering calculations[3,4] but they have their drawbacks. It is most important that the collision energy be kept appreciably lower than the potential energy of the three separate atoms, otherwise the scattering wave functions will have a significant amplitude in the regions of the potential where the reaction coordinates are not unique. It has also been noted that poor results have been obtained for systems such as IHI in which the central atom is much lighter than the end atoms because, from equation (4.12), the reactant and product asymptotic paths then subtend a narrow angle θ and the curvature of C will be very large near the saddle point. Finally, for three-dimensional systems the reaction coordinate approach becomes rather complicated as it is difficult to find a set of three coordinates which diagonalize the kinetic energy. These numerical problems seem to be avoided by using hyperspherical coordinates[5-7].

For the collinear triatomic system ABC, which is the simplest case, we start with the variables

$$x = \left(\frac{\mu_{A,BC}}{\mu_{BC}}\right)^{1/2} R_{A,BC}$$

$$y = r_{BC} \qquad 4.15$$

where μ are the reduced masses:

$$\mu_{BC} = m_B m_C/(m_B + m_C)$$

$$\mu_{A,BC} = m_A(m_B + m_C)/M \qquad 4.16$$

and $R_{A,BC}$ is the distance from A to the centre of mass of BC. The kinetic energy is diagonal in $\dot{x}$ and $\dot{y}$. The Delves polar coordinates are defined as follows:

$$r = (x^2 + y^2)^{1/2} \qquad 0 < r < \infty$$

$$\phi = \tan^{-1}(x/y) \qquad 0 < \phi < \phi_{max}$$

$$\phi_{max} = \tan^{-1}(m_B M/m_A m_C)^{1/2} \qquad 4.17$$

The origin of r is the point of coincidence of the three atoms. The reactant and product valleys are parallel to the lines $\phi = 0$ and $\phi = \phi_{max}$ as shown in Figure 4.3.

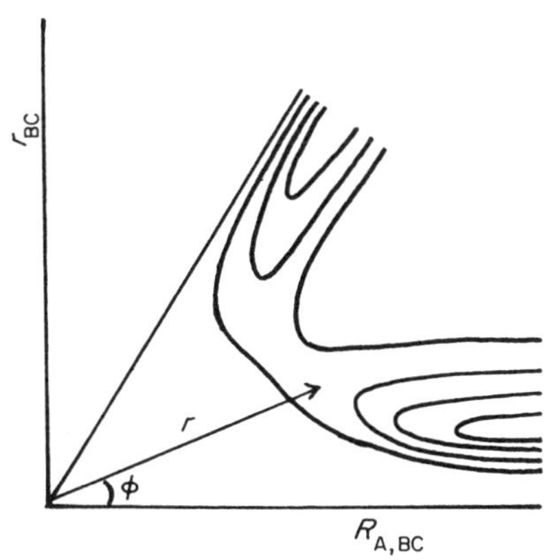

Figure 4.3 Delves polar coordinates (r, ϕ) for a collinear triatomic system

The Delves coordinates are suitable for describing the region of the surface in which the three atoms are far apart as well as the reactant and product valleys. They have been successfully used for situations in which the central atom is much lighter than the outer atoms[8,9]. Another advantage is that for symmetric atom exchange (A–B–A) the gerade and ungerade scattering wave functions can be propagated independently along the coordinate r, thus simplifying considerably the numerical calculations[10].

The potentials of tetra-atomic systems are functions of six internal coordinates and the six internuclear distances are a convenient set of these. The choice of spectroscopic coordinates depends on the equilibrium structure of the molecule. For example, the potential for a molecule such as NH_3 will generally be expressed in terms of three internuclear distances and three bond angles. Scattering coordinates will also depend on the system; different coordinates would be used for atom–triatom and diatom–diatom scattering.

For larger numbers of atoms one must likewise choose coordinates to suit the problem. It should be noted that for $n \geqslant 5$ the complete set of internuclear distances, numbering $\frac{1}{2}n(n-1)$ for n atoms, is greater than $3n-6$, the number of independent coordinates.

4.2 Symmetry coordinates

If a molecule contains two or more identical atoms its potential energy surface must be invariant to permutations of these atoms. Most chemists are familiar with the application of point group symmetry to problems of molecular structure and spectra, but they are less familiar with the uses of permutation groups. There are, however, few difficulties in making the transition from point to permutation groups and the most important permutation groups are isomorphic (have the same character table) with common point groups.

We shall start by considering a triatomic with three identical atoms such as O_3. The equilibrium structure of the ground state of this molecule has C_{2v} point group symmetry, the bond angle being 117°. In this geometry there are two types of atoms, a central atom and terminal atoms. It is clear that if the three oxygen nuclei were different isotopes and therefore distinguishable, then there would be three types of equilibrium molecule, as each of the three atoms could be central. A better way to express this is to say that the ground-state potential surface possesses three equivalent minima, each corresponding to a structure of C_{2v} symmetry.

If we set up an explicit function for the potential energy surface of O_3 in the coordinates R_1, R_2, R_3, which are even-handed with respect to the three atoms, then the potential must be invariant to permutations of these three coordinates. An equivalent statement is that the potential can contain only totally symmetric combinations of these three coordinates.

The permutation group of three, which is called S_3, has the elements

$$S_3\colon \{E, (123), (132), (12), (23), (31)\} \qquad 4.18$$

Table 4.1 Character tables for the permutation group S_3 and the point group C_{3v}.

S_3	E	(123) (132)	(12) (23) (31)	C_{3v}	E	$2C_3$	$3\sigma_v$
Γ_1	1	1	1	A_1	1	1	1
Γ_2	1	1	−1	A_2	1	1	−1
Γ_3	2	−1	0	E	2	−1	0

where, for example, (123) is a permutation operator for replacing 1 by 2, 2 by 3 and 3 by 1. The character table for S_3 is shown in Table 4.1. S_3 is isomorphic with the point group C_{3v} (also shown in Table 4.1), whose elements are

$$C_{3v}\colon \{E, C_3, C_3^2, \sigma_v, \sigma_v', \sigma_v''\} \tag{4.19}$$

where C_3 is a rotation by $2\pi/3$ about the principal axis and σ_v, σ_v' and σ_v'' are reflections in planes containing the principal axis.

If, for example, we wish to know the symmetries of the N–H stretching modes of NH_3 then we take a basis of displacement vectors along the three N—H bonds $(\Delta r_1, \Delta r_2, \Delta r_3)$; we show that this basis forms a representation of the group with characters

$$3, 0, 0, 1, 1, 1 \tag{4.20}$$

and further show that this has component irreducible representations A_1 and E. The combinations of basis functions which transform as these irreducible representations can be obtained by projection operator techniques as follows:

$$\begin{aligned} \psi_1(A_1) &= \sqrt{\tfrac{1}{3}}(\Delta r_1 + \Delta r_2 + \Delta r_3) \\ \psi_2(E) &= \sqrt{\tfrac{1}{2}}(\Delta r_2 - \Delta r_3) \\ \psi_3(E) &= \sqrt{\tfrac{1}{6}}(2\Delta r_1 - \Delta r_2 - \Delta r_3) \end{aligned} \tag{4.21}$$

These functions have been normalized and 'most common' forms of the E modes have been given.

The techniques for making the above analysis are described in any book on the chemical applications of group theory. We use exactly the same techniques to find those combinations of the variables (R_1, R_2, R_3) which transform as irreducible representations of the S_3 group. By strict analogy with the normal coordinate problem they are as follows:

$$\begin{aligned} S_1 &= \sqrt{\tfrac{1}{3}}(R_1 + R_2 + R_3) \\ S_2 &= \sqrt{\tfrac{1}{2}}(R_2 - R_3) \\ S_3 &= \sqrt{\tfrac{1}{6}}(2R_1 - R_2 - R_3) \end{aligned} \tag{4.22}$$

If the potential energy function is to be totally symmetric to permutation of the three variables, then it must be expressed in terms of S_1 or of totally symmetric combinations of S_2 and S_3. We know that there must be two, but only two, independent quadratic combinations of the basis that are totally symmetric. This is because the two combinations

$$\left.\begin{aligned} R_1^2+R_2^2+R_3^2 \\ R_1R_2+R_2R_3+R_3R_1 \end{aligned}\right\} \qquad 4.23$$

are obviously totally symmetric and no other totally symmetric functions can exist which are independent of these. In terms of the symmetry functions (4.22) the quadratic combinations are

$$\left.\begin{aligned} S_1^2 &= \tfrac{1}{3}(R_1^2+R_2^2+R_3^2+2R_1R_2+2R_2R_3+2R_3R_1) \\ S_2^2+S_3^2 &= \tfrac{2}{3}(R_1^2+R_2^2+R_3^2-R_1R_2-R_2R_3-R_3R_1) \end{aligned}\right\} \qquad 4.24$$

The appropriate combinations of S_2 and S_3 can be found by applying the projection operator for the Γ_1 irreducible representation to any of the quadratic combinations of S_2 and S_3.

Amongst the cubic functions, S_1^3 and $S_1(S_2^2+S_3^2)$ are obviously totally symmetric and there must be a third totally symmetric function [using a similar argument as before that three simple totally symmetric cubic combinations, analogous to (4.23), can be found]. It has a far from obvious form, namely

$$S_3^3-3S_3S_2^2 \qquad 4.25$$

This can also be found by applying the Γ_1 projection operator to cubic combinations of S_2 and S_3, although the algebra is rather tedious.

What about terms of higher order than cubic? Do we have to find at the quartic level another combination of S_2 and S_3 which is unrelated to those already used? The answer is no, because we can already construct four independent totally symmetric quartic terms, namely

$$S_1^4, \quad S_1^2(S_2^2+S_3^2), \quad S_1(S_3^3-3S_3S_2^2), \quad (S_2^2+S_3^2)^2 \qquad 4.26$$

In fact, there is an algebraic theorem which has been known for many years that all totally symmetric polynomials in three variables can be written as sums and products of three functions:

$$S_1, \quad S_2^2+S_3^2, \quad S_3^3-3S_3S_2^2 \qquad 4.27$$

and these three are called the integrity basis[11].The theorem can be generalized: for n variables all totally symmetric polynomials can be constructed from an integrity basis of n functions analogous to (4.27), one of each order up to n.

The reader might ask at this point what advantages there are in using symmetry variables instead of R_i variables. What advantages are there in working at the quadratic level, say, with (4.24) rather than (4.23)? The answer is none if all terms of a given order are to be used, but many if one is going to select terms of a given order and reject others of the same order. For example, if one is going to construct a completely general totally symmetric quartic polynomial then there will be four independent terms which can either be (4.26) or can be the set

$$\Sigma R_i^4, \quad \Sigma\Sigma R_i^3 R_j, \quad \Sigma\Sigma R_i^2 R_j^2, \quad \Sigma\Sigma\Sigma R_i^2 R_j R_k \tag{4.28}$$

Any function expressed in the set (4.26) can be equally well expressed in the set (4.28). However, if one has data to determine, say, only three of the four independent quartic terms, then there may be a preferred choice and, as we shall show in later examples, the use of symmetry coordinates has substantial advantages. It should be particularly noted that S_1 is a variable that defines the size of the three-atom triangle (it is proportional to the perimeter), whereas S_2 and S_3 are shape variables. It would be expected, therefore, that the potential energy function would show a different dependence to S_1 on the one hand and S_2 and S_3 on the other.

Let us turn to the more complicated example of an AX_4 molecule. If we label the four A–X distances R_1, R_2, R_3, R_4, then the potential energy function must be totally symmetric in the S_4 group which is isomorphic with the point group of a tetrahedron T_d. The basis R_1, R_2, R_3, R_4, can be transformed into functions that are a basis for irreducible representations of the group, and these have the same form as the normal modes for A–X stretch of a tetrahedral molecule. They are

$$\begin{bmatrix} S_1 \\ S_2 \\ S_3 \\ S_4 \end{bmatrix} = \frac{1}{2} \begin{bmatrix} 1 & 1 & 1 & 1 \\ 1 & 1 & -1 & -1 \\ 1 & -1 & 1 & -1 \\ 1 & -1 & -1 & 1 \end{bmatrix} \begin{bmatrix} R_1 \\ R_2 \\ R_3 \\ R_4 \end{bmatrix} \tag{4.29}$$

where S_1 is totally symmetric and (S_2, S_3, S_4) are components of a triply degenerate irreducible representation (T_1 in T_d). The integrity basis of these symmetry coordinates is[12]

$$S_1, \quad S_2^2+S_3^2+S_4^2, \quad S_2S_3S_4, \quad S_2^2S_3^2+S_3^2S_4^2+S_4^2S_2^2 \tag{4.30}$$

For the six X–X distances in an X_4 or AX_4 molecule the irreducible representations are even more complicated. Labelling the variables as in Figure 4.4, the transformation matrix is

$$\begin{bmatrix} S_1 \\ S_2 \\ S_3 \\ S_4 \\ S_5 \\ S_6 \end{bmatrix} = \begin{bmatrix} 6^{-1/2} & 6^{-1/2} & 6^{-1/2} & 6^{-1/2} & 6^{-1/2} & 6^{-1/2} \\ 2^{-1/2} & 0 & 0 & -2^{-1/2} & 0 & 0 \\ 0 & 2^{-1/2} & 0 & 0 & -2^{-1/2} & 0 \\ 0 & 0 & 2^{-1/2} & 0 & 0 & -2^{-1/2} \\ 0 & \frac{1}{2} & -\frac{1}{2} & 0 & \frac{1}{2} & -\frac{1}{2} \\ 3^{-1/2} & -12^{-1/2} & -12^{-1/2} & 3^{-1/2} & -12^{-1/2} & -12^{-1/2} \end{bmatrix} \begin{bmatrix} R_1 \\ R_2 \\ R_3 \\ R_4 \\ R_5 \\ R_6 \end{bmatrix} \tag{4.31}$$

Using the labels of the T_d irreducible representations, S_1 is A_1, (S_2, S_3, S_4) is T_2 and (S_5, S_6) is E. It should be noted that we are interested in the totally symmetric combinations of S_i under the permutations of the four X atoms (group S_4), not the more restrictive (integrity basis) set of totally symmetric combinations in the group S_6, as suggested in ref. 12.

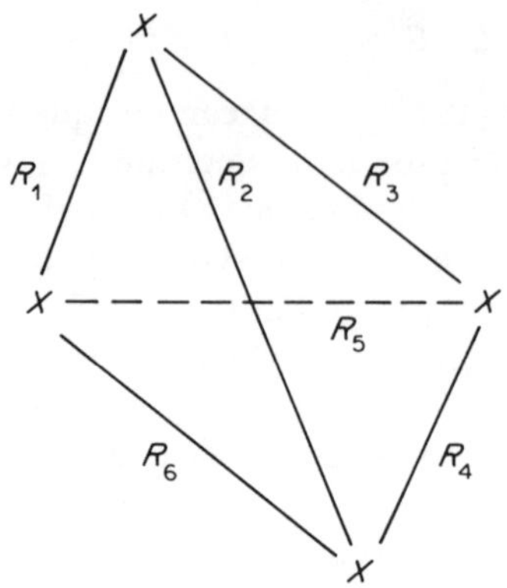

Figure 4.4 Labelling for the six X–X distances in an X_4 system

There are two totally symmetric quadratic forms, in addition to S_1^2, which are

$$S_2^2+S_3^2+S_4^2 \quad \text{and} \quad S_5^2+S_6^2 \tag{4.32}$$

Those cubic terms which cannot be written as products of lower-order terms can be obtained, one each, from the direct products $T_2 \times T_2 \times T_2$, $E \times E \times E$ and $E \times T_2 \times T_2$; this can be confirmed from the information in published tables of direct products (see Herzberg[13], Tables 32 and 33). The three combinations are as follows:

$$S_2S_3S_4$$

$$S_6^3 - 3S_6S_5^2$$

$$S_6(2S_2^2 - S_3^2 - S_4^2) + \sqrt{3}\,S_5(S_3^2 - S_4^2) \tag{4.33}$$

Explicit expressions have been given for the higher-order totally symmetric polynomials[14].

4.3 Optimized internal coordinates

An AB_2 triatomic molecule is the simplest system for which there is some advantage in using symmetry-adapted combinations of the internal coordinates to express the potential energy function. If R_1 and R_2 are the two A–B bond lengths, then

$$S_1 = \sqrt{\tfrac{1}{2}}(R_1 + R_2)$$

and

$$S_2 = \sqrt{\tfrac{1}{2}}(R_2 - R_1) \tag{4.34}$$

are the appropriate symmetry combinations, and the potential, being totally symmetric to exchange of the atoms B, can be expressed in terms of S_1 or S_2^2, but odd powers of S_2 are not to be used.

Let us now examine the relevance of this to a specific example HO_2. This

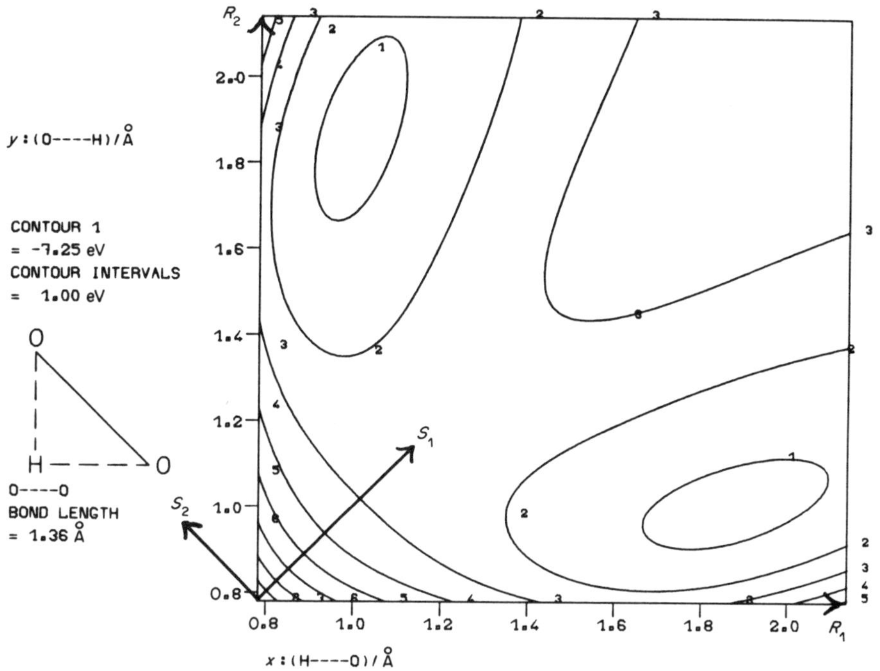

Figure 4.5 Contours of a potential for the ground state of HO_2 for a fixed O–O distance

molecule in its ground state has C_s symmetry with $H\hat{O}O = 106°$. However, there must be two minima on the surface corresponding to the equilibrium structure, which are related by exchange of the oxygen atoms. Figure 4.5 illustrates this through contours of a potential that has been deduced for the ground state of HO_2 (ref. 15). The contours are for varying R_1 and R_2 with the third variable as the O–O distance fixed at its equilibrium value of 1.36 Å.

The question we now raise is what is the simplest function that qualitatively represents the potential in the region local to the two minima? A reasonable answer to this is the function

$$V = (a + bS_1 + cS_1^2) + (dS_2^2 + eS_2^4) \qquad 4.35$$

It should be emphasized that, although one needs only a quadratic function in S_1, it is necessary to go to a quartic term in S_2 to obtain a two-minimum surface. Thus, although this approximate potential is quartic in both R_1 and R_2, all the quartic terms can be encompassed in S_2^4. The potential would certainly be improved by adding higher powers, including cross terms, in S_1 and S_2^2.

Figure 4.6 shows contours of a potential that has been deduced for the ground state of HCO (ref. 16) plotted in the same scheme as for Figure 4.5. This system again has two minima corresponding to the structures HCO and HOC, but in this case the minima are not symmetry related. By analogy with the previous discussion on HO_2, we would expect that the minimal function to represent the potential would be quartic in R_1 (R_{CH}) and R_3 (R_{OH}), but there are no symmetry

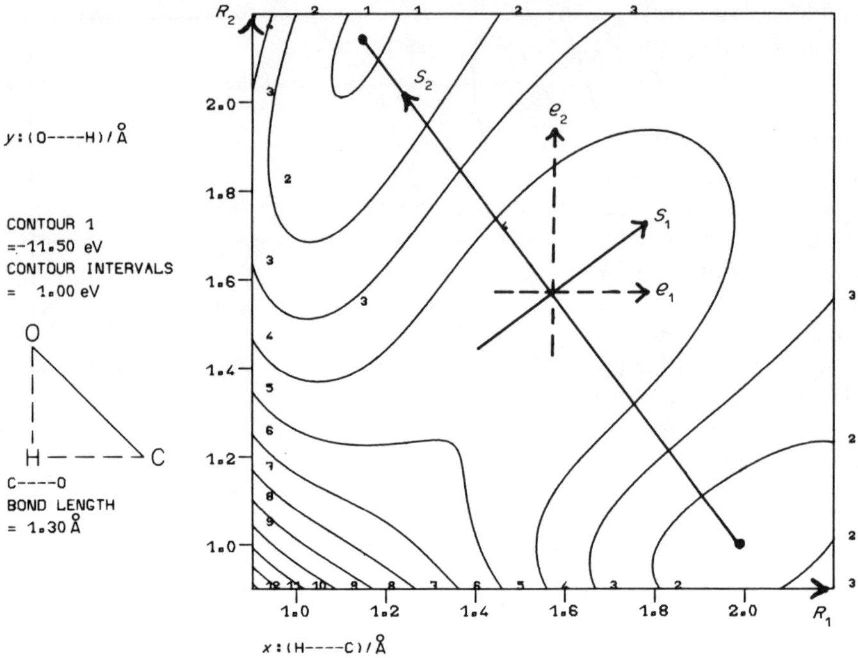

Figure 4.6 Contours of a ground-state potential for HCO for a fixed C–O distance, and the variables S_1 and S_2 which have been defined so that S_1 is zero at both minima on the surface

rules to guide us on whether all quartic terms are needed. In terms of the functions S_1 and S_2, defined by equations (4.34), the potential can contain odd powers of S_2. Progress in these circumstances can be made by transforming to new variables which are related to the coordinates of the minima on the surface.

Let the two minima on the surface of Figure 4.6 have coordinates (in R_i variables) (R_{1a}, R_{2a}) and (R_{1b}, R_{2b}). We first define new coordinates ρ_1 and ρ_2 by moving the origin to the mid point of the line joining these two minima:

$$\left.\begin{aligned}\rho_1 &= R_1 - \tfrac{1}{2}(R_{1a}+R_{1b})\\ \rho_2 &= R_2 - \tfrac{1}{2}(R_{2a}+R_{2b})\end{aligned}\right\} \qquad 4.36$$

We now rotate the axes so that new variables lie along the line of the minima and perpendicular to it, as shown in Figure 4.6. We shall again use the symbols S_1 and S_2 to represent these new variables. They are defined by

$$\left.\begin{aligned}S_1 &= \rho_1 \cos\alpha + \rho_2 \sin\alpha\\ S_2 &= -\rho_1 \sin\alpha + \rho_2 \cos\alpha\end{aligned}\right\} \qquad 4.37$$

where

$$\tan\alpha = \frac{R_{1b}-R_{1a}}{R_{2a}-R_{2b}} \qquad 4.38$$

By analogy with equation (4.35), it is now straightforward to write in S_i

variables a simple function that qualitatively represents the potential. We require a quadratic in S_1 and a quartic in S_2 as follows:

$$V=(a+bS_1^2)+(cS_2+dS_2^2+eS_2^3+fS_2^4) \quad 4.39$$

Of course, the potential can again be improved by further terms; the point we are emphasizing is that the minimal need is for higher powers in S_2 than in S_1.

Several triatomic systems that are of interest have three minima on their potential surfaces which may or may not be related by symmetry. For example, the ground state of SO_2 has at least three minima, one corresponding to the equilibrium C_{2v} structure and two that are symmetry equivalent that correspond to metastable C_s structures and the chemical formula SOO. It is possible to change from the (R_1, R_2, R_3) coordinate system by a translation followed by rotation to a system of new variables (S_1, S_2, S_3) which are optimally placed with respect to these three minima. The rules for carrying out these transformations are given in Appendix 2 and we shall use these rules several times in later chapters.

Having chosen optimum coordinates in order to fit the topography of a surface possessing several minima, we can look in more detail at the form of the polynomial representation of the surface around the minima. Expressions (4.35) and (4.39) show for two dimensions the simplest polynomials that must be used for the AB_2 and ABC two-minima surfaces. An alternative approach is to start with the spectroscopic or other data that may be available regarding these minima and to ask what polynomials could be determined.

Suppose, for example, we know the energy, equilibrium geometry and quadratic force field (second derivatives of the potential evaluated at the minima) for a system such as HO_2 which has two equivalent minima as illustrated in Figure 4.5. We have, in two dimensions, six pieces of data available, namely $(V)_e$, $(\partial V/\partial S_1)_e=0$, $(\partial V/\partial S_2)_e=0$, $(\partial^2 V/\partial S_1^2)_e=F_{11}$, $(\partial^2 V/\partial S_2^2)_e=F_{22}$ and $(\partial^2 V/\partial S_1\,\partial S_2)_e=F_{12}$, where the suffix e indicates evaluation at the equilibrium position. These data are sufficient to determine six parameters in a potential and, to be specific, six coefficients in a polynomial such as (4.35).

Now, equation (4.35) contains only five terms and we must first be certain that our data can determine the coefficients *a–e*. This can be confirmed by differentiating equation (4.35) and equating, as appropriate, to each of the above data except F_{12}. For the function (4.35) F_{12} would be zero and hence the knowledge of a non-zero value of F_{12} allows us to determine one more term in the polynomial. This term must involve both S_1 and S_2 and, by symmetry, must be an even power of S_2; the simplest such term is $S_1S_2^2$. The important point we are making is that the data available and the symmetry determine the polynomial that can be used to represent the surface. We could not, for example, determine the coefficients of S_1^3 from the available data, and any attempt to do so by setting up simultaneous equations linking the polynomial coefficients to data will be insoluble (singular matrices would arise in any matrix inversion procedure for solving such a problem).

Let us turn to the more complicated example of HCO and discuss the type of surface illustrated in Figure 4.6. In this case the energies, geometries and quadratic force fields of both minima will, in two dimensions, provide twelve pieces of data (twice that for HO_2). However, by the transformation of coordinates specified by equations (4.36) and (4.37) we have effectively satisfied two pieces of data, namely $(\partial V/\partial S_1)_e = 0$ at both minima, and this has been recognized by the non-appearance of an S_1 term in equation (4.39). The polynomial appearing in equation (4.39) has six unknown coefficients and therefore the data can, in principle, be used to determine these and a further four polynomial coefficients.

The data refer to minima whose coordinates are $(S_1, S_2) = (0, S_{2a})$ and $(0, -S_{2a})$ and it is easy to confirm that *a*–*d* in equation (4.39) plus a term in S_1S_2 can be determined by the data associated with any one of the two minima (but not $S_1S_2^2$ as in the case of HO_2 previously discussed); hence data on the second minimum can be used to determine *e*, *f* and another three terms. The energy of the second minimum will determine *e* (say) and $(\partial V/\partial S_2)_e = 0$ will determine *f* (say); the three quadratic derivatives will determine the coefficients of $S_1^2S_2$ from $(\partial^2 V/\partial S_1^2)_e$, $S_1S_2^2$ from $(\partial^2 V/\partial S_1\,\partial S_2)_e$ and S_2^5 from $(\partial^2 V/\partial S_2^2)_e$.

We emphasize that the available data will provide polynomial terms in S_2 up to fifth order and any attempt to omit this term and introduce instead a term like S_1^3 would lead to insoluble equations. By adopting optimum coordinates, as in Figure 4.6, this analysis is obvious, but it is not obvious in R_i space which polynomial terms to select.

The above analysis can be generalized to three dimensions and to systems possessing up to three minima on the surface which may or may not be equivalent by symmetry. This generalization is given in Appendix 2. Applications of the method will be described several times in later chapters, with the fullest account being in Chapter 10.

References

1. R. A. Marcus, *J. Chem. Phys.*, **45**, 4493, 4500 (1966).
2. R. A. Marcus, *J. Chem. Phys.*, **49**, 2610 (1968).
3. J. C. Light, *Methods Comput. Phys.*, **10**, 111 (1971), and references cited therein.
4. J. C. Light and R. B. Walker, *J. Chem. Phys.*, **65**, 1598 (1976).
5. A. Kuppermann, J. A. Kaye and J. P. Dwyer, *Chem. Phys. Lett.*, **74**, 257 (1980).
6. L. M. Delves, *Nucl. Phys.*, **9**, 391 (1959); **20**, 275 (1960).
7. V. Aquilanti, G. Grossi and A. Lagana, *Chem. Phys. Lett.*, **93**, 174 (1982).
8. J. A. Kaye and A. Kuppermann, *Chem. Phys. Lett.*, **77**, 573 (1981); **78**, 546 (1981).
9. J. Manz and J. Romelt, *Chem. Phys. Lett.*, **77**, 172 (1981).
10. G. Hauke, J. Manz and J. Romelt, *J. Chem. Phys.*, **73**, 5040 (1980).
11. H. Weyl, *The Classical Theory of Groups*, Princeton University Press, Princeton, NJ, 1946.
12. A. J. C. Varandas and J. N. Murrell, *Chem. Phys. Lett.*, **84**, 440 (1982).
13. G. Herzberg, *Infra-Red and Raman Spectroscopy of Polyatomic Molecules*, Van Nostrand, Princeton, NJ, 1945.
14. A. Schmelzer and J. N. Murrell, to be published.
15. S. C. Farantos, E. C. Leisegang, J. N. Murrell, K. Sorbie, J. J. C. Texeira-Dias and A. J. C. Varandas, *Mol. Phys.*, **34**, 947 (1977).
16. S. Carter, I. M. Mills and J. N. Murrell, *J. Chem. Soc., Faraday Trans. 2*, **74**, 148 (1978).

Chapter 5

Van der Waals molecules

5.1 Two-body interactions

The macroscopic properties of systems composed of closed-shell atoms or molecules can be explained fairly well by assuming that the total potential is the sum of the pair potentials in the system. This assumption is better for non-polar than for polar molecules, but even for polar molecules the assumption leads to qualitatively satisfactory results, as illustrated by the largely successful computer simulation of liquid water[1,2].

The validity of the pair-additive approximation can be examined most critically by *ab initio* calculations or spectroscopic studies on small clusters of atoms or molecules. Such clusters, which are known as van der Waals molecules, can be formed in relatively high concentrations in supersonic nozzle beams. In this chapter we examine in some detail the simplest type of van der Waals molecule, which is the cluster of inert gas atoms, and make general comments on other systems.

An approximate representation of the attractive region of the potential for an inert gas dimer can be obtained by combining an attractive term depending on R^{-6} and a repulsive term which is a rapidly rising function as R tends to zero. Two simple functions of this type have been given as equations (1.3) and (1.4). For an accurate representation of the potential one needs more elaborate functions and to understand their form it is necessary to give a brief account of the underlying theory[3,4].

If two systems A and B are sufficiently far apart that the exchange of electrons between them can be ignored (this is called the long-range region), then the set of all products of their wave functions

$$\psi_{rs}^{\mathrm{AB}} = \psi_r^{\mathrm{A}}(i)\psi_s^{\mathrm{B}}(j) \qquad 5.1$$

is a satisfactory basis for an expansion of the wave functions of the combined system AB. A perturbation development of the interaction gives, in first order, the electrostatic energy and, in second order, the polarization energies, which are a sum of two contributions, the induction energy and the dispersion energy.

For inert gas atoms in the long-range region the electrostatic energy and the induction energy (which is the energy arising from the polarization of one atom in the electrostatic field of the other) are both zero. The leading term in the perturbation expansion is therefore the dispersion energy which arises from the simultaneous polarization of the electron clouds around each atom. The dispersion energy itself can be expanded as a power series in $1/R$ and for closed shell systems the leading term is at R^{-6} [see expression (1.1)]. For practical purposes the series is terminated after the first, second or third term. The R^{-6} term arises from dipolar polarization on both atoms, R^{-8} from dipolar polarization on one atom and quadrupolar polarization on the other, etc. For the inert gas dimers reasonably accurate values have been deduced for the first three C_n coefficients[5].

When two atoms are separated by a distance which is approximately equal to the sum of their van der Waals radii or less it is no longer valid to ignore exchange of electrons between the two sets i and j, hence the product functions (5.1) are not a suitable basis for the system. Exchange is measured by the size of the overlap integrals between orbitals of A and orbitals of B, and in the region of the van der Waals minimum these integrals are small, although not negligible. This is referred to as the intermediate range of interactions. A satisfactory basis in this region can be obtained by antisymmetrizing the products (5.1):

$$\Psi_{rs}^{\mathrm{AB}} = \mathscr{A}\psi_r^{\mathrm{A}}(i)\psi_s^{\mathrm{B}}(j) \tag{5.2}$$

where $\mathscr{A}$ is the antisymmetrizing operator for the sets of electrons i and j.

The first-order energy now contains, in addition to the electrostatic energy, a term associated with the exchange of electrons between the two sets. The precise definition of this differs slightly from one perturbation method to another but the differences are not large. The feature which all theories possess is that the exchange energy is approximately proportional to the square of the largest overlap integral between orbitals of the two atoms. As overlap integrals decay exponentially to zero as the distance between the two centres increases the exchange energy will also show an exponential decay. The form of the short range repulsion in the function (1.4) is consistent with this, and so to a good approximation is any other rapidly changing function of R such as the R^{-12} term in the Lennard–Jones potential (1.3).

The exchange operator $\mathscr{A}$ will also influence the form of the long-range terms in the perturbation expansion: the electrostatic, induction and dispersion energies. Firstly, we note that although the electrostatic and induction energies are zero for inert gas atoms when overlap of orbitals is negligible, they are not zero in the intermediate range. However, these overlap-dependent contributions have, like the exchange energy, an exponential dependence on R and hence they can be subsumed into the exchange term. For the dispersion energy the situation is different. The R^{-n} expansion is valid only if there is no overlap of the polarized electron densities on the interacting systems and if there is such overlap then each term in this expansion overestimates the attraction arising from the relevant multipolar polarization. From studies which have been made on simple systems[6,7] it has been found that the C_n coefficients decrease as R decreases, and this is normally allowed for by introducing damping functions $D_n(R)$ according to the expression

$$C_n^*(R) = C_n D_n(R) \qquad 5.3$$

One attractive feature of this approach is that with an appropriate choice of damping functions the other contributions to the potential can be represented quite accurately by the Hartree–Fock energy of the system[8-10].

The most popular form of the damping function is[8,10,11]

$$D_n(R) = \exp[-\beta(1-\alpha R_m/R)^2], \quad R < \alpha R_m \qquad 5.4$$

where R_m is the van der Waals separation and α and β are parameters. For $R \geqslant \alpha R_m$, $D(R)$ is taken as unity. In early work the parameters α and β were taken to be independent of n, and values have been suggested by Ng *et al.*[11] ($\alpha = 1.28$, $\beta = 0.4$) which fit the potentials for all inert gas dimers.

More elaborate and accurate expressions for the damping functions have been derived by Koide *et al.*[12], Varandas and Brandão[13] and Douketis *et al.*[14]; in all studies different functions have been used for different n values. Koide *et al.*[12] derived damping functions of the form

$$D_n(R) = [1 - \exp(-A_n R - B_n R^2 - K_n R^3)]^n \qquad 5.5$$

from the *ab initio* potential of H_2 (b, $^3\Sigma_u^+$) for values of n from 6 to 20. The values of A_n, for example, are $0.3648a_0$, $0.3073a_0$ and $0.2514a_0$ for $n = 6$, 8 and 10, respectively. Varandas and Brandão[13] proposed a similar function to (5.5) but without the cubic term in the exponent. Their important contribution was to show that if R is replaced by R/ρ, where

$$\rho = (R_m + 2.5R_0)/2 \qquad 5.6$$

and

$$R_0 = 2(\langle r_A^2 \rangle^{1/2} + \langle r_B^2 \rangle^{1/2}) \qquad 5.7$$

$\langle r_A^2 \rangle$ and $\langle r_B^2 \rangle$ being the expectation values for the squared radii of the outer orbitals of the two atoms A and B, then the two constants A_n and B_n appearing in the exponential appear to have universal values for all van der Waals diatomics. Values of these constants for the first three dispersion terms are shown in Table 5.1. R_0 has been identified by Le Roy[15] as the minimum distance at which the undamped R^{-n} expansion may be considered a good representation of the total potential.

Table 5.1 Universal damping function coefficients derived by Varandas and Brandão[13].

	n		
Parameter	6	8	10
A_n	3.07191	2.23816	1.67176
B_n	8.82384	7.15284	5.97923

If one is interested only in an accurate representation of the potential without imposing any physical significance on the individual terms, then one has more flexibility in the functions that can be used. For example, it has been shown[16] that the function

$$V = A_0(1 + a_1 R + a_2 R^2 + a_3 R^3)\exp(-\alpha R) - \tanh(R - R_m/2)(C_6 R^{-6} + C_8 R^{-8} + C_{10} R^{-10}) \qquad 5.8$$

is an accurate representation of the whole potential for a very wide range of van der Waals systems. Note that the function $\tanh(R - R_m/2)$, which contains only the parameter R_m, cannot be strictly referred to as a damping function as it does not approach zero for small values of R. Instead, for $R < R_m/2$ the function becomes negative so that the $C_n R^{-n}$ terms make a positive contribution to the total potential.

A final comment should be made on the fact that if a correctly damped asymptotic expansion is used for the dispersion terms then the remainder of the potential can be obtained reasonably accurately from a Hartree–Fock calculation[17,18] or by exchange perturbation theory from Hartree–Fock atomic functions[19,20]. The reason for this is that the short-range terms depend primarily on the overlap of the electron densities of the two atoms, and for closed shell systems the Hartree–Fock density is, to first order, identical with the exact density, a consequence of Brillouin's theorem[21]. Additionally, it has been noted[19,20] that if one uses Hartree–Fock functions to construct the antisymmetrized basis (5.2) instead of exact functions, then a complementary exchange term, which appears in zeroth order of perturbation, is zero within the approximation of single exchange between the systems. Minimizing this zeroth-order exchange has been proposed as a criterion for calculating an accurate first-order exchange energy[22,23].

5.2 Three-body energies for the inert gases

The most important fact about the energy of inert gas trimers is that the three-body terms are very much smaller than the sum of the two-body energies in the intermediate and long ranges. That does not mean they can be ignored for all physical properties. For example, they have been found to contribute approximately 50% to the third virial coefficient of argon[24], two-body energies contributing the other 50%.

Following the pattern of the two-body energy one can separate the three-body energy into short- and long-range terms, the long-range terms being damped in the charge-overlap region.

The long-range three-body energies were first suggested by Axilrod and Teller[25] and independently by Muto[26]. Perturbation theory of three atoms without electron exchange gives, as the leading term, a triple-dipole dispersion energy which has the form

$$V^{(3)}_{\text{disp}} = C_9(1 + 3\cos\theta_1 \cos\theta_2 \cos\theta_3)/(R_1 R_2 R_3)^3 \qquad 5.9$$

where θ_i ($i=1, 3$) are the angles of the triangle formed by the atoms. Values of C_9 have been calculated for many systems[5]; although it is only the leading term in a $1/R$ expansion, higher terms are generally ignored. This triple-dipole term has been found to account for most of the three-body contribution to the third virial coefficient[24].

The three-body dispersion energies will be damped by charge overlap and this factor has been studied by O'Shea and Meath[27] for the lowest quartet state of H_3. Their results have been fitted by Varandas[28] to a functional form which is related to that of equation (5.6). There is the additional complication that the parameters must have an angular dependence, and R is replaced by the geometric mean of the three internuclear distances. The recommended universal form for all triple S-state interactions is

$$V^{(3)}_{\text{disp}} = C_9(R_1R_2R_3)^{-3}[1+84.72\exp(-10.84X)+3\cos\theta_1\cos\theta_2\cos\theta_3] \times[1-\exp\{-3.198(1-0.155\Omega_{123})X[1+1.507(1+0.688\Omega_{123})X]\}]^9 \tag{5.10}$$

where

$$\Omega_{123}=\sin\theta_1+\sin\theta_2+\sin\theta_3 \tag{5.11}$$

$$X^3=\rho_1\rho_2\rho_3 \tag{5.12}$$

$$\rho_i=R_i/R_{m,i};\quad i=1,2,3 \tag{5.13}$$

At the van der Waals equilibrium distance the damping factor causes a reduction of the order 5–10% of the Axilrod–Teller–Muto three-body term for the interaction of three H atoms with the same spin. Except for highly polarizable atoms (e.g. alkali metals), for which it is predicted to be much larger, that value should be typical.

Barker and Bobetic[29] have shown that with the best estimates of the Ar_2 potential and the *undamped* C_9 three-body term, the properties of solid argon are in good agreement with experiment. These properties include the nearest neighbour distance, the cohesive energy, the Debye parameter and the bulk modulus. Their calculations were, in fact, made with the inclusion of higher three-body dispersion terms (up to triple quadrupole) and with estimates of higher many-body dispersion terms, but, as shown by Bell and Zucker[30], these additional terms show a near cancellation. The calculations also suggest that there is a fortuitous cancellation of many-body overlap-dependent contributions to the argon crystal potential with the damped component of the C_9 three-body term.

Measures of the overlap-dependent contributions to the three-body energies of the inert gases can come from SCF calculations on the trimers[31] or from exchange perturbation theory[32]. These studies show that such terms are destabilizing for D_{3h} configurations and stabilizing, but very much smaller, for linear configurations. However, unless one is interested in internuclear distances that are appreciably less than van der Waals separations, these terms can safely be ignored. The only possible exception is for the long-standing problem of the relative stabilities of the face-centred cubic and hexagonal close-packed structures of the inert gas solids. Current theory favours an explanation for the very small extra stability of the fcc form based on overlap of the atomic wave functions[33,34], although a universally accepted form of the theory has probably yet to be produced.

5.3 Polyatomic van der Waals molecules

In this section we discuss briefly the potential energy functions for van der Waals molecules in which one or more of the components is a stable species. Examples are the HF dimer and the Ar–HCl complex. The new feature, which is absent for inert gas clusters, is that the potential energy function is almost separable into products of terms which represent the strong internal potential of the stable fragments and the weak external potentials describing the coupling between the fragments. Indeed, when we talk about the intermolecular potential between HF molecules (say), we usually mean the potential between two molecules rigidly fixed at their equilibrium distances. This will be a function of four coordinates $(x_1, \ldots, x_4)$ whose values define the separation and mutual orientation of the two diatomic species. These could, for example, be the distance between the centres of mass of the diatomic molecules and three angles defining their orientation, or the four interatomic distances.

On the assumption that the intermolecular potential between HF molecules is independent of the HF bond lengths (R_5 and R_6), we would write the total potential energy function as

$$V = V'(x_1, \ldots, x_4) + V_{\mathrm{HF}}(R_5) + V_{\mathrm{HF}}(R_6) \tag{5.14}$$

However, there will be some coupling of the internal and external terms, not least because the HF dipole moment depends on bond length and the dipole–dipole energy is the dominant long-range contribution to the intermolecular potential. For this reason it is convenient to write the potential for small displacements of R_5 and R_6 from their equilibrium values, in the form

$$V = V'(x_1, \ldots, x_4, R_5, R_6) + V_{\mathrm{HF}}(R_5) + V_{\mathrm{HF}}(R_6) \tag{5.15}$$

with the implication that V' is approximately equal to the V' in equation (5.14).

Whenever one has two dynamical processes, one of which is much faster than the other, it is possible to calculate the *faster* of the two at instantaneous positions of the particles in the *slower* motion. The *fast* motion then determines an effective potential for the *slow* motion. This is the basis of the Born–Oppenheimer separation of electron (the faster) and nuclear (the slower) motion. A similar separation can therefore be made for van der Waals molecules. The fast motion is now the motion of the nuclei within the stable fragments and the slow motion is the motion of the fragments against one another in an intermolecular potential which depends on the eigenstate of the fast motion. Again taking the HF dimer as our example, the intermolecular motion is determined by a potential

$$V'_{kl}(x_1, \ldots, x_4) \tag{5.16}$$

where k and l are the vibrational eigenstates of the HF molecules. V' in expression (5.16) is not equivalent to that in equation (5.15) because it is not a function of R_5 and R_6. It is also not identical with V' in equation (5.14) because of the state dependence, but it is very close to the V' in equation (5.14) because this state dependence is usually weak.

A further and similar decoupling of the rotational motion of the diatomic molecules from the motion along R, the distance between the two centres of mass, is also possible if the rotational motion is fast compared with the intermolecular motion (equivalently, if the rotational spacing is large compared with the spacing of quantized states for the R motion). A detailed account of this 'Born–Oppenheimer' approach to the dynamics of atom–diatom van der Waals molecules is given in ref. 35, but there are implied treatments in several earlier works on atom–diatom scattering.

There have been many studies of intermolecular potentials involving rigid molecules but little in the way of general strategy on the functional forms of these potentials. Even for an atom and a rigid diatomic molecule, which is only a two-variable problem, there are several options. The most popular approach is to use scattering coordinates (Figure 4.1b) and to write the potential as a Legendre polynomial expansion (4.2). However, it must be stressed that this expansion may be extremely slowly convergent unless the diatomic molecule is nearly spherical.

The problem arises mainly from the repulsive region of the potential the onset of which may be very sensitive to θ. It is clear that the expansion could not be used for the interaction between a hard-sphere atom and a hard-ellipsoid diatomic molecule and, in so far as repulsive potentials can be (very roughly) approximated by hard interactions, one can see that this is the source of the problem.

An alternative to the expansion (4.2) that has been used by several workers replaces R by a dimensionless variable which is scaled to the value of R at the van der Waals minimum; this is a function of θ. Writing

$$\eta = [R - R_m(\theta)]/R_m(\theta) \qquad 5.17$$

the expansion has the general form

$$V(\eta, \theta) = \sum_{\lambda} F_\lambda(\eta) P_\lambda(\cos\theta) \qquad 5.18$$

The function $V(0, \theta)$, for example, is the potential at the bottom of the van der Waals minimum as the atom moves around the diatomic molecule. This can usually be expressed accurately by a sum of very few Legendre polynomials.

An alternative approach to atom–diatom potentials which obviates the Legendre expansion is to use a sum of two central potentials which are functions of the internuclear distances. Thus for a complex AB–C the potential will be the sum of a function of R_{AC} and one of R_{BC}, both functions being normal atom–atom potentials: Morse, Lennard–Jones or better. This cannot, in general, be an exact representation of the potential, but it can be improved either by adding further (off-nuclear) central potentials or by making the two atom–atom potentials have an angular dependence. However, neither of these extensions has been widely employed.

For a diatom–diatom potential there is a similar choice, either an expansion in angular variables (necessarily bipolar because both diatomic molecules make angles with the distance R between the centres of mass), or the sum of four atom–atom central potentials. For polar diatomics an important component of the long-range potential is the electrostatic energy, and in the bipolar expansion the terms are the well-known multipole interactions: dipole–dipole, dipole–quadrupole, etc. However, it now appears to be firmly established that the convergence of the multipolar expansion is frequently slow and a representation of the electrostatic energy based on distributed charges[36] or atom-centred multipoles is more accurate[37].

The only intermolecular potential of the atom–rigid diatom that we discuss in detail is that between an inert gas and a hydrogen halide. The experimental information on these systems is substantial, consisting of the rotational spectra of the complexes formed in nozzle beams, pressure broadening of the hydrogen halide rotational lines, virial coefficients and some scattering data. References to much of this work are contained in two theoretical papers by Hutson and Howard[38,39].

The potential is based on a generalized Lennard–Jones function (1.3), used by Maitland and Smith[40], in which the parameters depend on the scattering angle variable θ. It takes the form

$$V(R,\theta)=\varepsilon(\theta)\left(\frac{6}{n-6}x^{-n}-\frac{n}{n-6}x^{-6}\right) \tag{5.19}$$

where

$$\left.\begin{aligned} x&=R/R_m(\theta)\\ n&=m(\theta)+\gamma(x-1)\end{aligned}\right\} \tag{5.20}$$

The angle-dependent variables are written

$$\begin{aligned} \varepsilon(\theta)&=\varepsilon(0^\circ)+\Delta\varepsilon_1 f(\theta;\lambda)+\Delta\varepsilon_2 g(\theta)\\ R_m(\theta)&=R_m(0^\circ)+\Delta R_1 f(\theta;\kappa)+\Delta R_2 g(\theta)\\ m(\theta)&=m(0^\circ)+\Delta m f(\theta;\xi) \end{aligned} \tag{5.21}$$

where

$$f(\theta;\lambda)=1-[\tfrac{1}{2}(1+\cos\lambda)]^{\lambda} \tag{5.22}$$

The function f increases monotonically from a value of zero at 0° to unity at 180° and the rate of change is controlled by the parameters λ, κ and ε.

The function $g(\theta)$ is defined by

$$\left.\begin{aligned} g(\theta)&=\cos^2\theta \quad \theta>90^\circ\\ g(\theta)&=0; \qquad\quad \theta<90^\circ\end{aligned}\right\} \tag{5.23}$$

and this term was included because of evidence from a preliminary analysis of the rotational spectra of NeHCl and NeDCl (ref. 38) that there were two minima on the potential surface. The deeper was linear with the H(D) inside, and the second, shallower, represented a collinear structure with the halogen on the inside. There is supporting evidence for the secondary minimum in this type of complex from scattering cross-sections of Ar and HCl (ref. 41) and from *ab initio* calculations on Ar–HCl (ref. 42).

The potential defined by equations (5.19)–(5.23), which Hutson and Howard refer to as *M*5, contains twelve parameters and optimization of these, mainly to spectroscopic data for the inert gas–HCl complexes, gave the results shown in Table 5.2[38]. Contour plots are shown in Figure 5.1. The similarity of all the contour plots is striking, and if the energies are scaled to $\varepsilon(0°)$ and the distances to $R_m(0°)$ then it can be shown that they are a conformal family (i.e. all closely represented by the same potential in scaled coordinates).

Table 5.2 Parameters of the *M*5 potential determined by Hutson and Howard[38]; tolerances on these parameters are quoted in the original work.

Parameter	NeHCl	ArHCl	KrHCl	XeHCl
$\varepsilon(0°)/\text{cm}^{-1}$	68.22	180.51	213.86	263.61
$\Delta\varepsilon_1/\text{cm}^{-1}$	−28.72	−79.35	−97.05	−127.60
$\Delta\varepsilon_2/\text{cm}^{-1}$	14.0	34.0	39.0	43.0
λ	4.62	4.04	4.61	4.66
$R_m(0°)/\text{Å}$	3.76	4.01	4.10	4.27
$\Delta R_1/\text{Å}$	−0.09	−0.12	−0.09	−0.11
$\Delta R_2/\text{Å}$	−0.16	−0.16	−0.16	−0.16
κ	12.7	12.11	12.7	12.7
$m(0°)$	22.2	22.39	22.2	22.2
Δm	−9.6	−9.93	−9.6	−9.6
ξ	16.4	17.25	16.4	16.4
γ	9.0	9.0	9.0	9.0

This example of the inert gas–hydrogen halide intermolecular potential has been given in some detail because it represents the level of sophistication that can be achieved at present by fitting an analytical function to empirical data. Spectroscopic data are inevitably more sensitive to the region of the potential around the minimum and in some complexes of this family, particularly the HF complexes[39], it has been realized that the spectroscopic data provide little evidence for the form of the potential in the region of the secondary minimum, or even whether it exists or not. *Ab initio* calculations would in these cases be a more reliable guide to the potential in such regions.

Finally, one should also note that the angular variables used in this work, (5.24) and (5.25), are not standard Legendre functions and hence the potential is unlikely to be suitable for all atom–diatom molecule potentials.

At present the only van der Waals systems for which there has been an attempt to obtain the coupling terms between the inter- and intramolecular motion from empirical data are the inert gas–H_2 complexes. One implication of the 'Born–

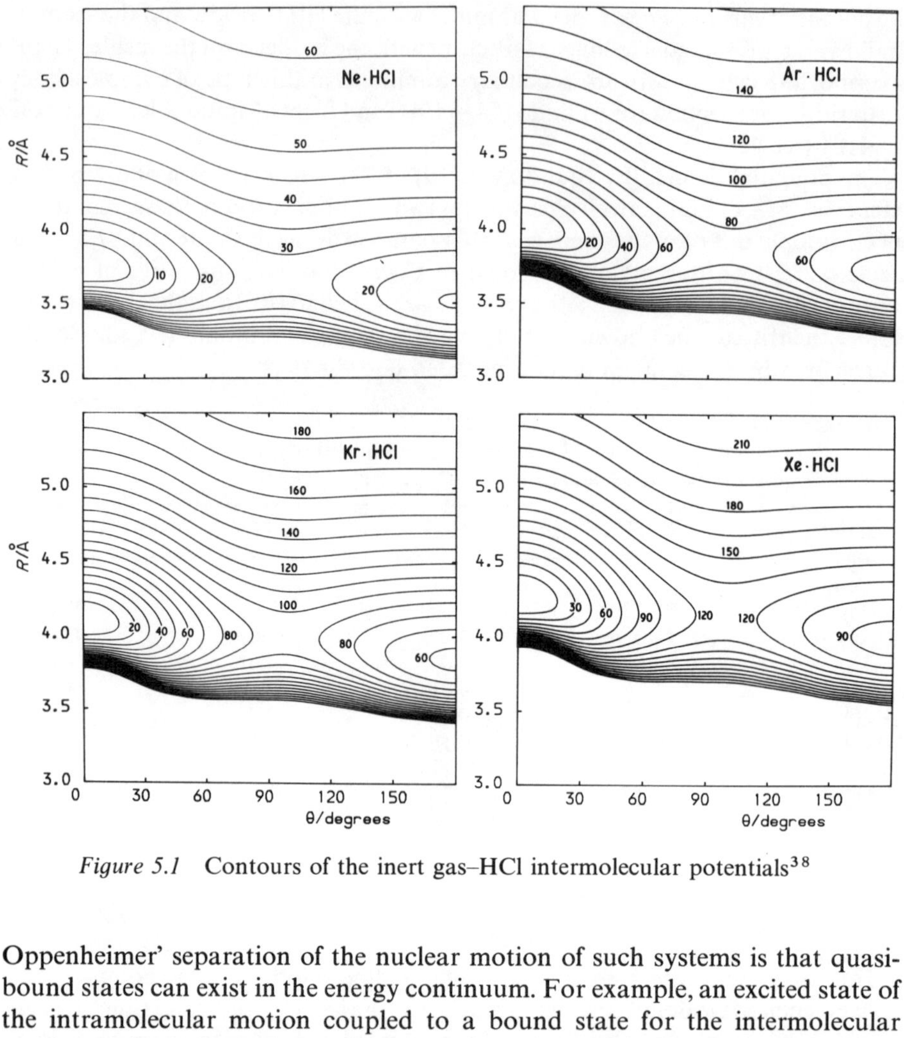

Figure 5.1 Contours of the inert gas–HCl intermolecular potentials[38]

Oppenheimer' separation of the nuclear motion of such systems is that quasi-bound states can exist in the energy continuum. For example, an excited state of the intramolecular motion coupled to a bound state for the intermolecular motion can have the same energy as the intramolecular ground state coupled to a continuum state of the intermolecular motion. This is illustrated in Figure 5.2 for a model AB–C system.

The breakdown of the Born–Oppenheimer approximation will give rise to predissociation of the intramolecular excited state and line broadening of the molecular spectra. This phenomenon, which was predicted theoretically before it was observed experimentally (for a review of early work, see ref. 43), has been extensively analysed by Le Roy *et al.* to determine accurate potentials for the inert gas–H_2 complexes[44].

A potential was constructed of the form

$$V(R, \xi, \theta) = \sum_{k=0}^{3} \sum_{\lambda=0,2} \xi^k P_\lambda(\cos\theta) V_{\lambda k}(R) \qquad 5.24$$

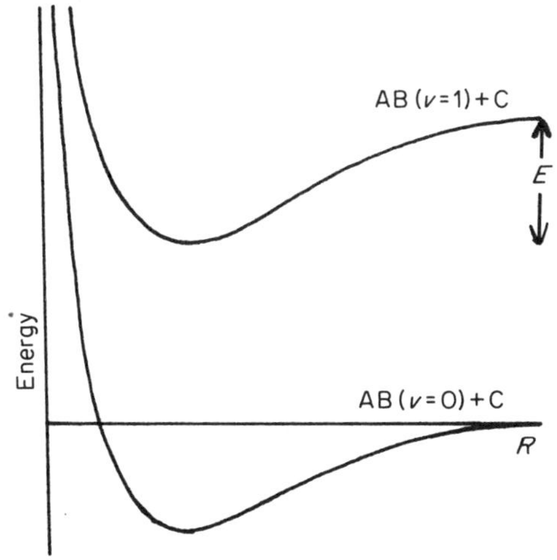

Figure 5.2 Effective potentials for an AB–C system for the ground and first excited vibrational states of AB. R is the distance from C to the centre of mass of AB. The energy range shown as E is in the continuum of the ground state, but spans the bound states of the $v=1$ potential

where R is the distance between the inert gas atom and the H_2 centre of mass, ξ is a dimensionless displacement of the H–H bond length from its equilibrium value $[\xi=(r-r_0)/r_0]$ and θ is the angle between R and the H–H internuclear axis. By symmetry, only even Legendre functions can appear in this expansion. The radial functions were given the form

$$V_{\lambda k}(R)=A^{\lambda k}\exp(-\beta_\lambda R)-D(R)[C_6^{\lambda k}R^{-6}+C_8^{\lambda k}R^{-8}] \quad 5.25$$

where

$$D(R)=\exp[-4(R_m/R)-1]^3 \quad \text{for } R\leqslant R_m$$
$$D(R)=1 \quad \text{for } R>R_m \quad 5.26$$

R_m is the position of the minimum of the function V_{∞}. Note that this function has a damped dispersion term analogous to (5.4) and an exponential repulsion.

Parameters for the H_2–Ar, –Kr and –Xe potentials have been derived by Carley and Le Roy[45] and vibrational averages of these parameters for the $(v, J)=(0, 0)$, $(0, 2)$, $(1, 0)$ and $(1, 2)$ H_2 states of the H_2–Ar complex have been calculated[44]. It has been deduced that the depth of the van der Waals minimum is 50.68 cm^{-1} in the (0, 0) state but 52.31 cm^{-1} in the (1, 2) state.

Given the success achieved for diatomic van der Waals systems by separating the Hartree–Fock and correlation energies it would be expected that this approach would be successful also for polyatomic interactions if combined with the many-body expansion. Varandas[46] has examined this procedure for the He–H_2 system. The Hartree–Fock components of the two- and three-body terms were obtained from accurate SCF calculations, and the correlation components were obtained semi-empirically, by interpolation from the dispersion coefficients of the various separate and united atom limits (i.e. He–He, H–Li, H–H–He) and those also known for He–H_2 (equilibrium). This semi-empirical potential compares well with the most accurate *ab initio* calculations on the He–H_2 surface[47], and requires no more data than a popular model of Tang and Toennies[48], which assumes a Legendre expansion of the potential.

To complete this chapter, we restate the point that intermolecular potentials, even in the adiabatic representation in which the potential depends on the internal states, are valid only for small displacements of the internal coordinates from their equilibrium values. If, for example, one requires a potential for the HF dimer which is valid also in the $H_2 + F_2$ valley, then it is necessary to treat it as for strongly coupled atoms, and apply the procedure that will be described in Chapter 12. The many-body expansion, as used in ref. 46 and in another, less accurate, treatment of He–H_2 (ref. 49), is a useful approach to van der Waals potentials.

References

1. O. Matsuoka, E. Clementi and M. Yoshimine, *J. Chem. Phys.*, **64**, 1351 (1976).
2. G. C. Lie, E. Clementi and M. Yoshimine, *J. Chem. Phys.*, **64**, 2314 (1976).
3. B. Pullman (Ed.), *Intermolecular Interactions from Diatomics to Biopolymers*, Wiley, New York, 1978.
4. M. L. Klein and J. A. Venables (Eds.), *Rare Gas Solids*, Vol. 1, Academic Press, London, 1976, Ch. 2 and 3.
5. K. T. Tang, J. M. Norbeck and P. R. Certain, *J. Chem. Phys.*, **64**, 3063 (1976).
6. J. N. Murrell and G. Shaw, *J. Chem. Phys.*, **49**, 4731 (1968).
7. H. Kreek and W. J. Meath, *J. Chem. Phys.*, **50**, 2289 (1969).
8. J. Hepburn, G. Scoles and R. Penco, *Chem. Phys. Lett.*, **36**, 451 (1975).
9. K. T. Tang and J. P. Toennies, *J. Chem. Phys.*, **66**, 1496 (1977); **67**, 375 (1977); **68**, 786 (1978).
10. R. Ahlrichs, R. Penco and G. Scoles, *Chem. Phys.*, **19**, 119 (1977).
11. K. C. Ng, W. J. Meath and A. R. Allnatt, *Mol. Phys.*, **37**, 237 (1979).
12. A. Koide, W. J. Meath and A. R. Allnatt, *Chem. Phys.*, **58**, 105 (1981).
13. A. J. C. Varandas and J. Brandão, *Mol. Phys.*, **45**, 857 (1982).
14. C. Douketis, G. Scoles, S. Mordetti, M. Zen and A. J. Thakkar, *J. Chem. Phys.*, **76**, 3057 (1982).
15. R. J. Le Roy, Chemical Society Specialist Periodical Report, *Molecular Spectroscopy*, Vol. 1, Chemical Society, London, 1973, p. 113.
16. P. Huxley, D. B. Knowles, J. N. Murrell and J. D. Watts, *J. Chem. Soc., Faraday Trans. 2*, in press.
17. G. Das and A. C. Wahl, *Phys. Rev., Sect. A*, **4**, 825 (1971).
18. J. P. Toennies, *Chem. Phys. Lett.*, **20**, 238 (1973).
19. A. Conway and J. N. Murrell, *Mol. Phys.*, **27**, 873 (1974).
20. J. N. Murrell and A. J. C. Varandas, *Mol. Phys.*, **30**, 223 (1975).
21. L. Brillouin, *Actual. Sci. Ind.*, **71** (1933); **159** (1934).
22. A. J. C. Varandas, *Chem. Phys. Lett.*, **69**, 222 (1980).
23. A. T. Amos and C. S. van den Berghe, *Mol. Phys.*, **47**, 897 (1982).
24. C. H. J. Johnson and T. H. Spurling, *Aust. J. Chem.*, **24**, 2205 (1971).
25. B. M. Axilrod and E. Teller, *J. Chem. Phys.*, **11**, 299 (1943).
26. Y. Muto, *Proc. Phys. Math. Soc. Jpn.*, **17**, 629 (1943).
27. S. F. O'Shea and W. J. Meath, *Mol. Phys.*, **28**, 1431 (1974).
28. A. J. C. Varandas, *Mol. Phys.*, **49**, 817 (1983).

29. J. A. Barker and M. V. Bobetic, *J. Chem. Phys.*, **79**, 6306 (1983).
30. R. J. Bell and I. J. Zucker, in *Rare Gas Solids*, Eds. M. L. Klein and J. A. Venables, Academic Press, London, 1976, Vol. 1, Ch. 2.
31. O. Novaro and F. Nieves, *J. Chem. Phys.*, **65**, 1109 (1976), and references cited therein.
32. D. R. Williams, L. J. Schaad and J. N. Murrell, *J. Chem. Phys.*, **47**, 4916 (1967).
33. K. F. Niebel and J. A. Venables, in *Rare Gas Solids*, Eds. M. L. Klein and J. A. Venables, Academic Press, London, 1976, Vol. 1, Ch. 9.
34. S. Carter and J. N. Murrell, *Mol. Phys.*, **32**, 1269 (1976).
35. S. L. Holmgren, M. Waldman and W. Klemperer, *J. Chem. Phys.*, **67**, 4414 (1977).
36. J. T. Brobjer and J. N. Murrell, *J. Chem. Soc., Faraday Trans. 2*, **78**, 1353 (1982).
37. A. J. Stone, *Chem. Phys. Lett.*, **83**, 233 (1981).
38. J. M. Hutson and B. J. Howard, *Mol. Phys.*, **45**, 769 (1982).
39. J. M. Hutson and B. J. Howard, *Mol. Phys.*, **45**, 791 (1982).
40. G. C. Maitland and E. B. Smith, *Chem. Phys. Lett.*, **22**, 443 (1973).
41. U. Buck and J. Schlensener, *J. Chem. Phys.*, **75**, 2470 (1981).
42. T. van Dam, A. Rozendaal, A. Vliegenthart and F. B. van Duijneveldt, to be published.
43. J. A. Beswick and J. Jortner, *Adv. Chem. Phys.*, **47**, 363 (1981).
44. R. J. Le Roy, G. C. Covey and J. M. Hutson, *Faraday Discuss. Chem. Soc.*, **73**, 339 (1982), and references cited therein.
45. J. S. Carley and R. J. Le Roy, *Adv. Chem. Phys.*, **42**, 353 (1980).
46. A. J. C. Varandas, to be published.
47. W. Meyer, P. Hariharan and W. Kutzelnigg, *J. Chem. Phys.*, **73**, 1880 (1982), and references cited therein.
48. K. T. Tang and J. P. Toennies, *J. Chem. Phys.*, **68**, 5501 (1976); **74**, 1148 (1981); **76**, 2524 (1982).
49. S. C. Farantos, J. N. Murrell and S. Carter, to be published.

Chapter 6

Fitting to ab initio *calculations*

6.1 A least-squares surface for LiFH

The first chemical reaction to be studied by crossed molecular beams was K+HBr, by Taylor and Datz in 1955[1]. In the following years many other examples of the reaction between hydrogen halides and alkali metals were examined and they are of considerable interest to theoreticians because their characteristics vary markedly with the different metal or halogen used[2]. The most important features of such beam reactions are the distribution of products over the scattering angle, the amounts of vibrational and rotational energy in the product alkali halide and the effect on the reaction cross-section of internal energy in the reactant hydrogen halide. All of these features should be interpretable, for the most part, by classical dynamical studies of the reactions, using accurate potential energy surfaces. Quantum mechanical studies of reactive scattering are at present very difficult but will no doubt be feasible before too long, and these should lead to a complete interpretation of the scattering experiment in terms of the potential surface function.

For this general class of reaction the system LiFH is the most amenable to accurate *ab initio* calculation of the surface because it has the least number of electrons, and it has therefore received considerable attention by quantum chemists. Lester and Krauss[3] in 1970 carried out SCF calculations on those regions of the surface relevant to non-reactive collisions between Li and HF and confirmed the presence of a shallow well associated with the hydrogen bonded structure Li···H—F which had earlier been postulated by Herschbach[4]. At a later date two semi-empirical valence bond studies were published, by Balint-Kurti and Yardley[5] and by Zeiri and Shapiro[6], in which the whole surface was examined. There were, however, significant differences in the surface features found in these studies, the most important being that the transition state was found in the exit channel of the reaction (6.1) by Balint-Kurti and Yardley (i.e. towards the LiF+H asymptote), but in the entrance channel by Zeiri and Shapiro.

$$\mathrm{Li} + \mathrm{HF} \rightarrow \mathrm{LiF} + \mathrm{H} \qquad 6.1$$

It is well known from classical dynamical studies that the vibrational energy which is found in the products, and the effect on the reaction of vibrational energy in the reactants, are very sensitive to the position of the transition state[7].

Chen and Schaefer[8] were stimulated to make a thorough study of the LiFH surface both by the disagreement in the valence-bond results and by more extensive experimental work being undertaken by Becker *et al.*[9]. They used the *ab initio* SCF MO technique with a large basis including polarization functions. The main weakness in their work by present standards is that configurations were generated only by single and double excitations from a single root function, namely the SCF wave function. For greater accuracy one would today generate configurations by excitation from several root functions, all those which have a substantial weight in the final CI wave function. Chen and Schaefer suggested that because of this restriction the calculated energy of their transition state was too high by about 0.1 eV (2 kcal mol^{-1}) relative to the energy of the reactants.

The *ab initio* calculations support the prediction of Balint-Kurti and Yardley[5] that the transition state for reaction (6.1) is non-linear and lies towards the LiF + H asymptote. The transition state is reached after passing through a shallow minimum representing a weakly bound complex which also has a bent configuration.

Chen and Schaefer reported energies of 214 points on the potential surface. However, about half of these are clustered around the geometries of the transition state and the complex and the other half provide a rather sparse coverage of the surface. Our attempts[10] to obtain a potential function by least-squares fitting to the *ab initio* points gave functions which showed an unphysical minimum in the LiF + H exit valley with $\mathrm{Li\hat{F}H} \approx 45°$ and to prevent this a few interpolated points in this region were added to the data.

A 0.1 eV error in the height of the barrier along a reaction coordinate may not seem large, and is indeed small by the general standards of good *ab initio* calculations. However, such an error can have a large effect on calculated scattering cross-sections or reaction rate constants. To produce a surface which is going to be useful for the interpretation of experimental data it is therefore necessary to modify the barrier to give agreement with the best experimental or theoretical estimate. This was achieved by subtracting from the *ab initio* energies amounts which followed a gaussian distribution centred at the transition state of height 0.1 eV. The width of this distribution is speculative without further theoretical evidence, but appears to give a physically acceptable surface.

Another unsatisfactory feature of the *ab initio* results is that they are not in exact agreement with experimental data pertaining to the diatomic fragments. The calculated equilibrium bond lengths of both LiF and HF are slightly too long and the exothermicity of the reaction is slightly underestimated. To make the *ab initio* points merge smoothly with the empirical potential energy curves of the diatomic fragments at their equilibrium distances, 0.041 Å was subtracted from all values of R_{LiF} and 0.016 Å from all values of R_{HF}. To make a smooth interpolation to the empirical dissociation energies of the diatomic fragments an amount 0.03 eV was subtracted at the LiF + H asymptote, this interpolating smoothly to zero at short Li–H or H–F distances[10].

It must be emphasized that these corrections to the *ab initio* results which are intended to make the surface conform to the best experimental or theoretical estimates are speculative because there is no firm evidence of how errors vary over the surface. In Section 9.3 we examine an alternative approach to the treatment of such errors which may be more satisfactory but which requires a knowledge of the *ab initio* potentials of the diatomic species.

The application of the Wigner–Witmer rules to the system LiFH shows that the ground-state surface, which is a doublet state, dissociates adiabatically to the ground states of all three atom–diatom products

$$\mathrm{LiFH}(\tilde{X},{}^2A')\rightarrow \begin{array}{l}\mathrm{HF}(X,{}^1\Sigma^+)+\mathrm{Li}(^2\mathrm{S})\\ \mathrm{LiF}(X,{}^1\Sigma^+)+\mathrm{H}(^2\mathrm{S})\\ \mathrm{LiH}(X,{}^1\Sigma^+)+\mathrm{F}(^2\mathrm{P})\end{array} \qquad 6.2$$

and it also dissociates to give three atoms in their ground states

$$\mathrm{LiFH}(\tilde{X},{}^2A')\rightarrow \mathrm{H}(^2\mathrm{S})+\mathrm{Li}(^2\mathrm{S})+\mathrm{F}(^2\mathrm{P}) \qquad 6.3$$

There are, of course, other surfaces of LiFH which give the same dissociation products. For example, the spins of three doublet states can be coupled to give two doublet and one quartet state. Also, the $F(^2P)$ state is three-fold spatially degenerate, so on interaction with $LiH(X, {}^1\Sigma^+)$ it will give one A'' and two A' surfaces. Taking the plane of the LiFH nuclei to be the x, y plane, P_x and P_y are symmetric to reflection and P_z is antisymmetric to reflection in this plane.

A further complication is that spin–orbit coupling in the fluorine atom splits its ground state into two components, ${}^2P_{3/2}$ (the lower) and ${}^2P_{1/2}$ (the upper). This is a first-order splitting, because there is spatial degeneracy in the P state, although it is very small (404 cm^{-1}). In $LiFH(\tilde{X}, {}^2A')$ spin–orbit coupling is quenched; more precisely, it becomes second order because the spin–orbit coupling Hamiltonian mixes doublet states of different energies. The ground-state surface of LiFH dissociates adiabatically to give $F(^2P_{3/2})$ atoms. This fact is most important if one is interested in the collision dynamics of LiH and fluorine atoms. If the fluorine atoms are cold (i.e. all in the ${}^2P_{3/2}$ state) then there is a 50% chance (there are four components of ${}^2P_{3/2}$ and two components of ${}^2A'$) of the collision occurring on the $LiFH(\tilde{X}, {}^2A')$ surface. If the fluorine atoms are hot (equal population of ${}^2P_{3/2}$ and ${}^2P_{1/2}$) then there is a $33\frac{1}{3}$% chance of the collision occurring on the $LiFH(\tilde{X}, {}^2A')$ surface.

Carter and Murrell[10] have derived a many-body expansion of the LiFH potential from the adjusted *ab-initio* data, neglecting spin-orbit coupling. The three-body contribution to the potential was obtained by subtracting the empirical two-body potentials from the adjusted *ab initio* energies (3.22):

$$V^{(3)}_{\mathrm{LiFH}}(R_1,R_2,R_3)=V_{\mathrm{LiFH}}-V^{(2)}_{\mathrm{HF}}(R_1)-V^{(2)}_{\mathrm{LiF}}(R_2)-V^{(2)}_{\mathrm{LiH}}(R_3) \qquad 6.4$$

The three-body term contained a complete quartic polynomial in the three internuclear distances:

$$P = V^0\left(1 + \sum_i C_i\rho_i + \sum_{i\leq j} C_{ij}\rho_i\rho_j + \cdots\right) \tag{6.5}$$

where $\rho_i = R_i - R_i^0$, and a range function of the form (3.26):

$$T = \prod_{i=1}^{3} [1 - \tanh(\gamma_i\rho_i/2)] \tag{6.6}$$

There are 41 parameters in all to be optimized: V^0, 34 coefficients in P, three values of R_i^0 and three of γ_i. This was achieved by a least-squares fit to the 214 calculated and four interpolated points on the surface.

Of the parameters appearing in equations (6.5) and (6.6), V^0 and the coefficients C have a linear relationship with the right-hand side of equation (6.4) and the least-squares procedure applied to these can be solved by matrix inversion. The other parameters, γ_i and R_i^0, are non-linear and an indirect, and therefore time-consuming, minimization procedure has to be adopted for which good initial guesses are desirable.

As a starting point for the minimization, the values $R_i^0 = 1$ Å and $\gamma_i = 2$ Å^{-1} were selected. These choices were made by experience with other three-body terms. The values of R_i^0 can be thought of as defining a reference structure to which the complete surface is related. If there is one feature of a surface of special importance (e.g. the minimum on a single minimum surface) then there are usually advantages to be gained by using this as the reference. If there is no such point then our experience is that the reference should be some point which is approximately central to the most important features of the surface. The optimum values of γ_i are similar to the γ which appear in the two-body terms (1.19), although no precise relationship to these has been found.

Although the optimization of a 41 parameter function, non-linear in some parameters, would appear to be a formidable task, no difficulty was found in this example in reaching the optimum function which is given in Table 6.1. Although there is always a risk that such optimizations will lead to a local minimum rather than the true minimum, the program that was used tested for this possibility by optimizing from different initial conditions and no change was found in the outcome. The optimum function gave a standard deviation to the data points of 0.08 eV (1.8 kcal mol^{-1}) with a maximum error of 0.29 eV. This maximum error occurred at a rather high point on the surface and hence does not have much physical significance. The height of the saddle point was fitted to better than 0.01 eV.

Figure 6.1 shows the contours of the potential for $\mathrm{Li\hat{F}H} = 114°$. The minimum that is shown corresponds to the complex in the entrance channel of the forward reaction (6.1) as predicted by *ab initio* calculations. The potential function defined by the parameters in Table 6.1 gives the depth of this minimum as 0.26 eV relative to HF + Li and its geometry is $R_{\mathrm{HF}} = 0.93$ Å, $R_{\mathrm{FLi}} = 1.90$ Å, $\mathrm{Li\hat{F}H} = 114°$. Figure 6.2 shows similar contours for LiFH = 74°, and includes the transition state at $R_{\mathrm{HF}} = 1.31$ Å, $R_{\mathrm{FLi}} = 1.65$ Å whose energy is 0.34 eV above the asymptote. Finally, Figure 6.3 shows contours for collinear F—H—Li which

Table 6.1 Parameters of the LiFH potential. The two-body terms are Extended Rydberg potentials (1.19):

$$V^{(2)} = -D_e(1 + a_1\rho + a_2\rho^2 + a_3\rho^3)\exp(-a_1\rho), \quad \rho = R - R_e$$

The three-body term is defined by equations (6.5) and (6.6). R_1, R_2 and R_3 are the H–F, Li–F and Li–H distances, respectively.

Two-body terms:

Species	D_e/eV	R_e/Å	a_1/Å^{-1}	a_2/Å^{-2}	a_3/Å^{-3}
HF	6.1173	0.9168	4.0604	3.3165	3.0396
LiF	5.9457	1.5639	2.0221	0.7432	0.9094
LiH	2.5161	1.5953	2.1822	1.1094	0.4576

Three-body term:

C_1/Å^{-1}	−1.8971	C_{111}/Å^{-3}	−9.7549	C_{1111}/Å^{-4}	3.6954	C_{1123}/Å^{-4}	0.2290
C_2	0.3390	C_{222}	0.2230	C_{2222}	−0.0917	C_{1223}	0.1840
C_3	0.0774	C_{333}	0.0822	C_{3333}	−0.1050	C_{1233}	−0.3421
		C_{112}	1.5351	C_{1112}	−0.0472		
C_{11}/Å^{-2}	10.0494	C_{122}	0.2157	C_{1122}	0.3544	V^0/eV	12.1649
C_{22}	0.0972	C_{113}	3.2750	C_{1222}	−0.0769		
C_{33}	−0.0658	C_{133}	0.1000	C_{1113}	−2.0080	γ_1/Å^{-1}	2.5058
C_{12}	−0.4097	C_{223}	−0.4650	C_{1133}	−0.3388	γ_2/Å^{-1}	1.1114
C_{13}	−1.6688	C_{233}	0.2617	C_{1333}	−0.3155	γ_3/Å^{-1}	0.1406
C_{23}	−0.1330	C_{123}	−0.5099	C_{2223}	0.1368		
				C_{2233}	−0.0582	R_1^0/Å	0.4687
				C_{2333}	0.0962	R_2^0	1.6211
						R_3^0	1.7020

includes the weak hydrogen-bonded structure whose existence was first suggested by Herschbach[4]. The minimum is very shallow, only approximately 0.3 kcal mol^{-1}, but we have confirmed that the complex is stable to angle bending.

Quantum dynamics in the collinear mode[11] and full quasi-classical trajectories[12] have been obtained for reaction (6.1) on the potential energy surface described above, and the results have been compared with the experimental molecular beam observations[9].

The quasi-classical trajectories show that reaction proceeds above a translational energy of 3 kcal mol^{-1}, as found experimentally, although at this low energy the calculated cross-section is only about a third of the experimental value. This is probably due to the failure of the classical trajectories to simulate tunnelling. At the higher translational energy of 8.7 kcal mol^{-1} the calculated cross-section of 1.03 ± 0.08 Å^2 was in excellent agreement with the experimental value, as was the prediction of forward peaking in the scattering angle. The results were much more satisfactory than those obtained with an earlier semi-empirical surface[13,14], and it has been concluded[12] that the surface defined in Table 6.1 is accurate when tested by the available experimental evidence.

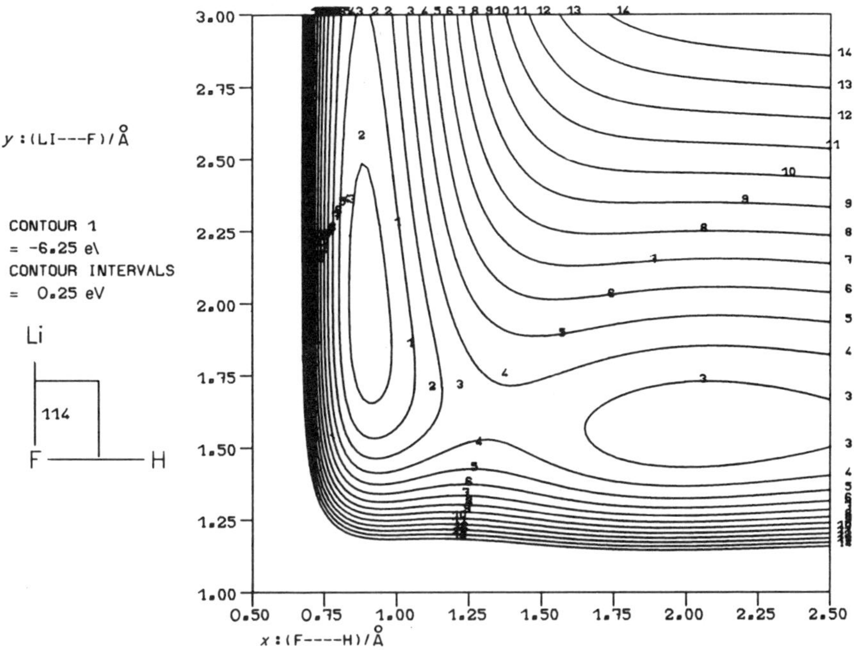

Figure 6.1 Contours of the LiFH potential defined by the parameters of Table 6.1 for LiF̂H = 114°

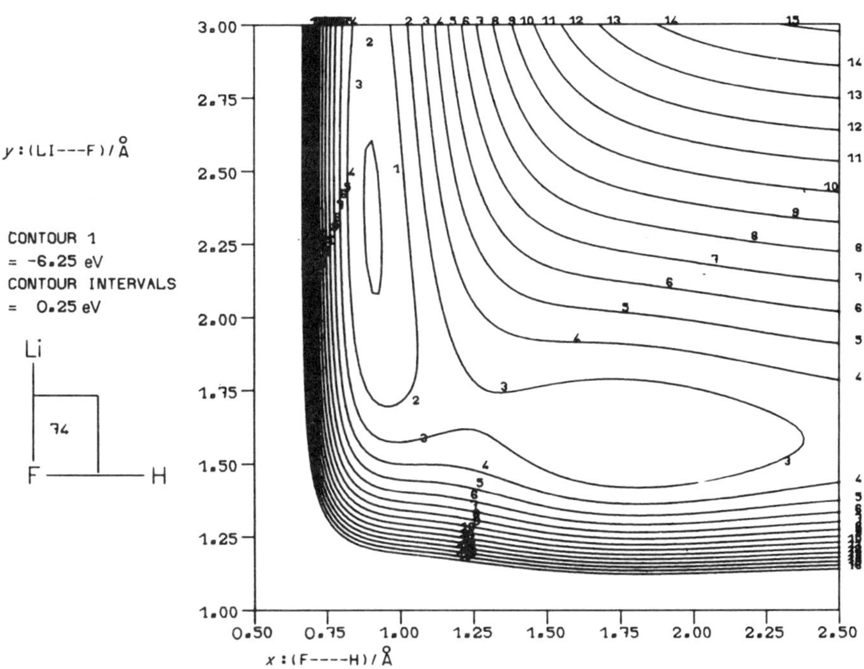

Figure 6.2 Contours of the LiFH potential for LiF̂H = 74°. The saddle point in this figure is the transition state for reaction (6.1)

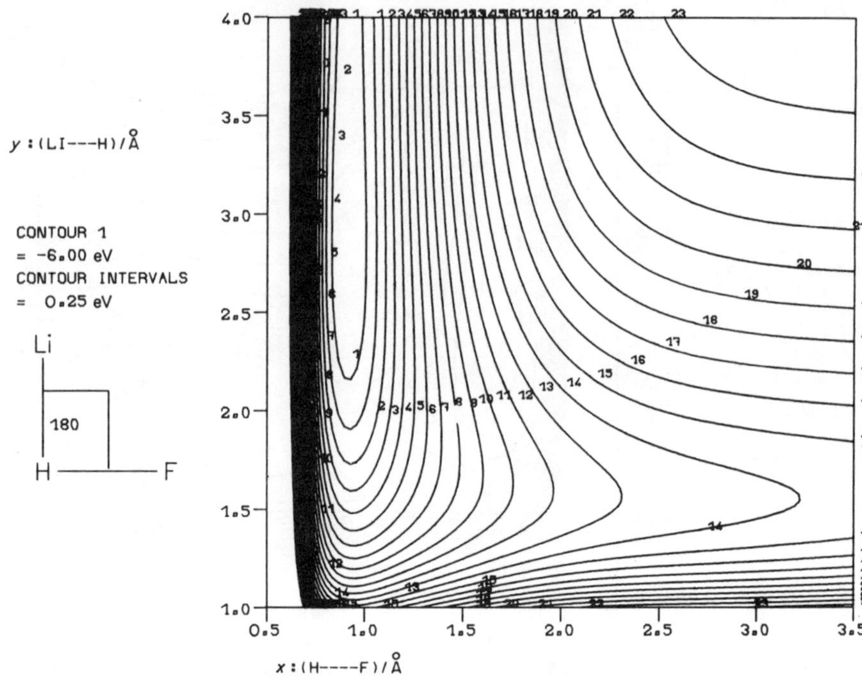

Figure 6.3 Contours of the LiFH potential for LiF$\hat{H}$ = 180°

6.2 The alkali metal trimers

This section illustrates the use of the quantum mechanically based LEPS function (Section 3.2) to obtain a potential energy surface. In contrast to the previous example in which a multiparameter function was optimized to a large number of *ab initio* points on the surface, the LEPS function has few parameters and these are chosen to reproduce some key features of the surface such as the positions and energies of stationary points. The example is placed in this chapter because these positions and energies were determined from *ab initio* calculations, but for other systems wholly or partly experimental data may be used to determine the LEPS parameters.

The alkali metal trimers are some of the simplest systems for which potential surfaces have been studied by quantum mechanical calculations. As the atoms have one valence electron in an s orbital, the electronic states of the trimer have some similarities with those of H_3, which is discussed in detail in Chapter 8. However, the important difference from H_3 is that they are stable molecules, although only for Li_3 has the dissociation energy been determined— and that with poor accuracy[15]. The systems Na_3, K_3 and Cs_3 have been observed in molecular beam studies[16] and the ESR spectra of Na_3 and K_3 (ref. 17) have further indicated that they are chemical species rather than just van der Waals complexes. Further support for this is the fact that the cross-sections for exchange reactions of the type

$$X' + X_2 \rightarrow X + XX'$$

and

$$Y + X_2 \rightarrow X + XY$$

are large, about 100 Å^2 (refs. 18–21).

Accurate *ab initio* calculations including correlation energy are available for the ground states of Li_3 (refs. 22–26) and Na_3 (ref. 27). An important conclusion of these studies is that the energy is relatively insensitive to variation of the bond angle in the trimer. As it is difficult to devise a basis which is equally favourable for all geometries an accurate prediction of the optimum geometry for these systems is very difficult. The ground states of the symmetric alkali metal trimers have a Jahn–Teller cusp at D_{3h} symmetries, like H_3, and there is evidence from calculations for a crossing along a line in the surface for the unsymmetrical system LiNaK[28]. However, the equilibrium geometries of Li_3 and Na_3 are only slightly displaced from D_{3h} structures, the predicted bond angles being close to 70° in both cases[22-27]. The state symmetry is $^2B_2$ in the C_{2v} point group.

The simplest quantum mechanical description of the surfaces for three ^{2}S atoms is provided by the LEPS function (Section 3.2) but the simplest analysis with this function, using Morse (1.2) and anti-Morse functions for the singlet and triplet states of the diatomic fragments, leads to the prediction that the lowest energy structure is linear[29]. Companion[30] was the first to note that by including the van der Waals well in the diatomic triplet state potential the lowest energy structure is predicted to be bent; an angle of 101° was found for Li_3. Using one of the more flexible derivatives of the LEPS approach (see Section 3.2) there is therefore some hope that one can obtain an accurate representation of these surfaces. This has been confirmed by the work of Varandas and Morais[31].

It is worth reflecting at this point that the differences between H_3 and the alkali metal trimers must arise from the influence of the low-lying vacant p orbitals for the latter. These p orbitals will contribute to directional bonding and, perhaps more important, they will increase the depth of the van der Waals minimum in the triplet state because low-energy s to p excitations will increase the dispersion energy.

For the single-state potentials Varandas and Morais used the Hybrid Extended Rydberg function[32]:

$$^1V = -D_e\left[1 + \sum_{i=1}^{5} a_i(R - R_e)^i\right]\exp\left[-a_1(R - R_e)\right]; \quad R \leqslant R_0 \qquad 6.7$$

$$^1V = -\sum_{L=3}^{5} C_{2L}R^{-2L} - BR^{\beta}\exp(-\alpha R); \qquad R > R_0 \qquad 6.8$$

where R_0 is given by expression (5.7). For the triplet states they used the sum of V_{HF}, the Hartree–Fock potential, and a damped dispersion term defined as in equations (5.5)–(5.7):

$$^3V = {^3V_{HF}} - \sum_{L=3}^{5} C_{2L} R^{-2L}\{1 - \exp[-d_1^{(2L)}x(1 + d_2^{(2L)}x)]\}^{2L} \quad 6.9$$

$$x = R\rho^{-1} \quad 6.10$$

$$\rho = (R_m + \gamma R_0)/2 \quad 6.11$$

The parameters $d_1^{(2L)}$ and $d_2^{(2L)}$ fill the roles of A_n and B_n in expression (5.5), and can be obtained from values given in Table 5.1. The numerical values of the other parameters appearing in equations (6.7)–(6.11) are given in Table 6.2.

Table 6.2 Constants used to define the diatomic potentials in the semi-empirical LEPS calculations for the alkali metal trimers[31]; all values are in atomic units.

Parameter	Li_2	Na_2	K_2	Rb_2	Cs_2
$10^2 D_e$	3.8513	2.6827	1.9109	1.8117	1.6647
R_e	5.0512	5.8179	7.4134	7.7989	8.4376
$10a_1$	7.9174	7.9642	7.4447	7.4182	6.4064
$10a_2$	1.0079	1.1244	1.1074	1.2944	0.7210
$-10^3 a_3$	6.2081	9.1555	8.8490	2.2497	1.7420
$-10^3 a_4$	2.6819	3.0527	2.3615	4.5438	1.7711
$10^4 a_5$	2.6983	3.4357	3.2199	4.6929	1.6132
$10^{-3} C_6$	1.39	1.51	3.89	4.87	7.26
$10^{-6} C_8$	0.0840	0.112	0.446	0.610	1.15
$10^{-8} C_{10}$	0.0722	0.107	0.549	0.802	1.72
$10^2 B$	1.3555	1.250	0.300	0.212	0.107
β	4.556	4.591	5.173	5.295	5.391
α	1.260	1.252	1.134	1.112	1.072
$10^{-1} R_0$*	1.69	1.82	2.00	2.10	2.30
$10^{-1} R_m$†	0.825	0.948	1.086	1.119	1.180

* The values of R_0 actually used to obtain ρ from equation (6.11) were slightly different from those given here and were, in the above order, 16.91, 17.56, 20.32, 21.04 and 22.76; see reference 33.
† Location of the van der Waals minimum for the lowest triplet state of the alkali metal dimers; see equation (6.11).

The Hartree–Fock potentials of Li_2, LiNa and Na_2 can all be fitted by a screened Coulomb potential of the form[33]

$$^3V_{HF} = aR^{-1} \exp[-b(R - R_e)] \quad 6.12$$

where R_e is the singlet state equilibrium distance. The accuracy of the fit can be seen from Figure 6.4. Varandas and Brandão[33] suggested that the parameters determined from this figure, $a = 0.0830 E_h a_0$ and $b = 0.7125 a_0^{-1}$, provide a universal triplet Hartree–Fock potential for all alkali dimers.

The Coulomb and exchange integrals, Q and A, respectively, that appear in the LEPS expression (3.11), can be expressed as analytical functions of the internuclear distance by using the Heitler–London expression (3.4) (with $S = 0$).

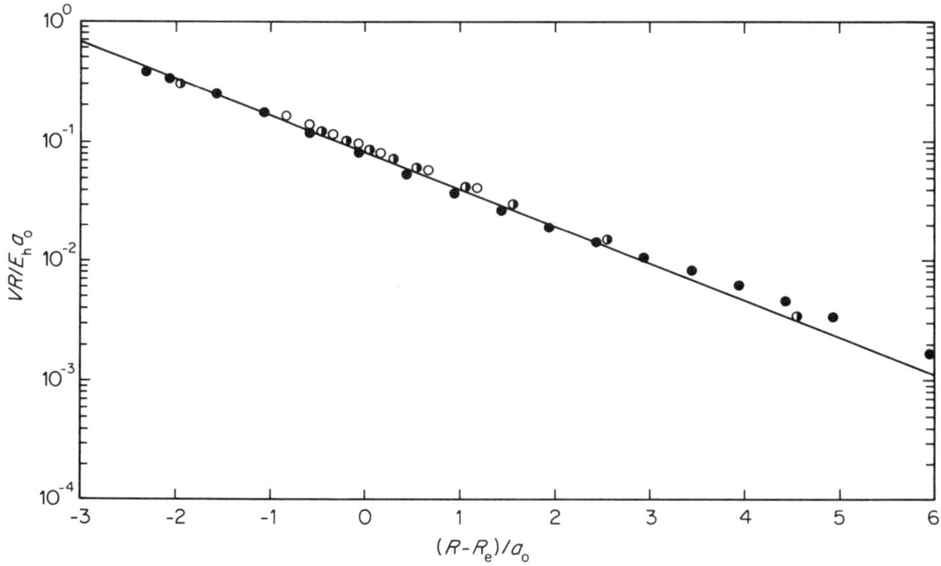

Figure 6.4 Comparison of the near Hartree–Fock potentials of Li_2 (●), LiNa (◑) and Na_2 (○), with the best screened Coulomb potential of the form (6.12) (full line)

The only undefined parameter in the calculation is γ, which appears in expression (6.11). For the triplet states of the dimers it has been found that $\gamma = 2.5$ gives the best description of the potential but if we consider γ to be optimum for the trimer, in order to account for some ionic contributions to the wave function, then a value $\gamma = 2.05$ is found to reproduce most exactly the Li_3 experimental dissociation energy and *ab initio* geometry. The same value also gives a good representation of the Na_3 *ab initio* geometry and it has been proposed as a universal constant for all such systems. The predicted equilibrium structures of Li_3 and Na_3 obtained by this procedure are shown in Table 6.3, as also are the

Table 6.3 Comparison of *ab initio* energies and geometries with those obtained from the semi-empirical LEPS potential[31]. ΔE is the stabilization energy relative to the atom plus diatom limit. R is the bond length and θ the bond angle. The *ab initio* data are from references 23, 25 and 27.

		Ab initio			Semi-empirical[31]		
Trimer	Symmetry	R/a_0	$\theta/°$	$\Delta E/E_h$	R/a_0	$\theta/°$	$\Delta E/E_h$
Li_3	$^2B_2$	5.2	71	0.0161	5.2	74	0.0269
	$^2A_1$	5.7	54	0.0150	5.7	52	0.0264
	$^2E'$	5.7	60	0.0128	5.4	60	0.0247
	$^2\Sigma_u^+$	5.5	180	0.0085	5.2	180	0.0220
Na_3	$^2B_2$	6.2	73	0.0136	6.0	77	0.0161
	$^2A_1$	6.4	52	0.0126	6.7	51	0.0155
	$^2E'$	6.5	60	0.0110	6.3	60	0.0136
	$^2\Sigma_u^+$	6.4	180	0.0076	6.0	180	0.0136

optimum energies and bond lengths of the D_{3h} ($^2E'$), $D_{\infty h}$ ($^2\Sigma_u^+$) and C_{2v} ($^2A_1$) structures. The optimum D_{3h} geometry is bound relative to atom plus ground-state diatomic and is doubly degenerate. Distortion along an e mode ($+q$) to give an obtuse triangle geometry leads to the equilibrium structure, $^2B_2$. The second sheet of the two-valued surface is destabilized in this direction (cf. Figure 2.1). However, distortion in the opposite direction ($-q$) stabilizes the second sheet to give a minimum energy structure of $^2A_1$ symmetry which is actually a saddle point on the ground-state surface.

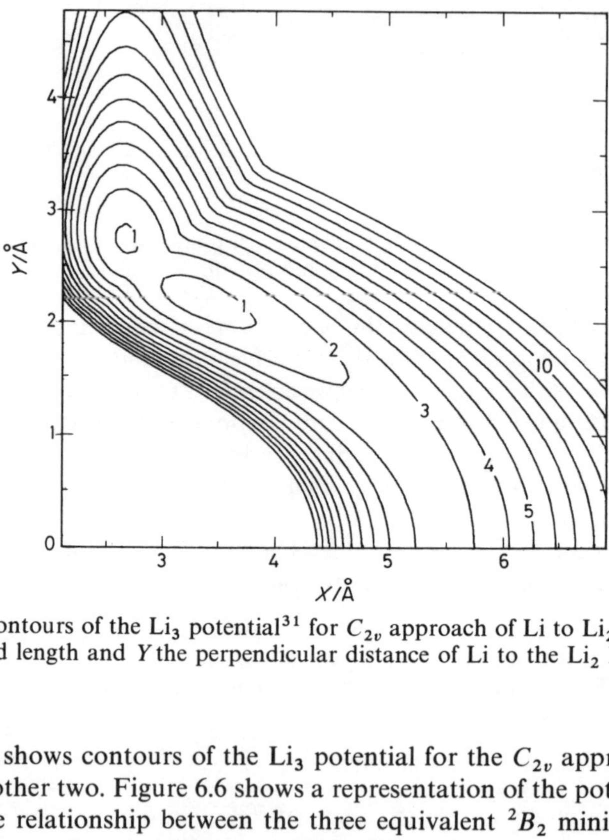

Figure 6.5 Contours of the Li_3 potential[31] for C_{2v} approach of Li to Li_2. X is the Li_2 bond length and Y the perpendicular distance of Li to the Li_2 line

Figure 6.5 shows contours of the Li_3 potential for the C_{2v} approach of one atom to the other two. Figure 6.6 shows a representation of the potential which illustrates the relationship between the three equivalent $^2B_2$ minima and $^2A_1$ saddle points on the ground-state surface. Contours are drawn for a fixed perimeter of the Li_3 triangle of 8.83 Å which is the equilibrium perimeter for the $^2B_2$ structure. The coordinates of such a triangular graph are explained in more detail in Section 7.4 (see Figure 7.8).

It can be seen from Table 6.2 that the barriers to interconversion of the $^2B_2$ equilibrium structures are very low (approximately 0.4 kcal mol^{-1}) and the molecule behaves dynamically as if it were a D_{3h} molecule[25]. The semi-empirical potentials suggest that this is true for all the alkali metal trimers.

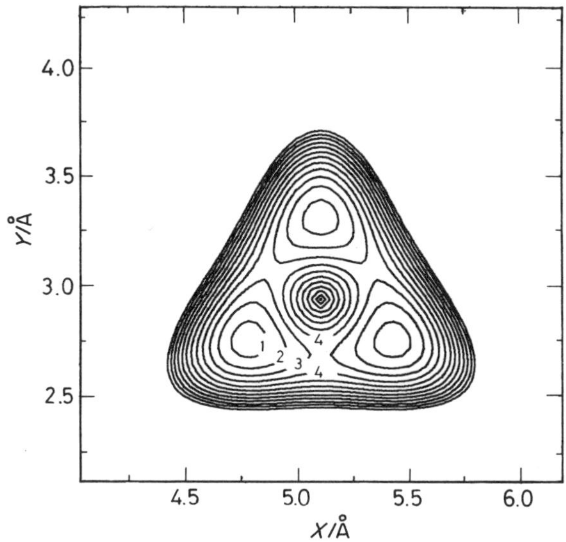

Figure 6.6 Contours of the Li_3 potential[31] shown as a triangular graph for a fixed perimeter of the Li_3 triangle of 8.83 Å

6.3 The H_2F surface

We conclude this chapter with a third example of a surface which has been extensively studied by quantum mechanical methods, and the data used to obtain potential energy functions. The H_2F surface has probably received more attention from theoreticians than any other, with the possible exception of H_3, the interest arising from the experimental work on the exothermic reaction

$$F + H_2 \rightarrow FH + H \quad 6.13$$

The potential energy release in this reaction is approximately 32 kcal mol^{-1}, which is sufficient to produce HF molecules in vibrational states up to $v = 3$. The vibrational distribution produced in the reaction departs strongly from an equilibrium distribution, and infrared chemiluminescence and a chemical laser can be produced[34,35]. The reaction has also been studied in crossed molecular beams[36].

The first extensive *ab initio* calculation on the HF surface was by Bender *et al.*[37]. They used a basis of double-zeta quality and a CI expansion with about 200 configurations. Most of their work was focused on linear FHH structures (150 points on the surface), another 200 points being calculated for bent structures with the F$\hat{H}$H bond angle up to 90°. These calculations confirmed earlier speculation that the transition state for reaction (6.13) is linear. The predicted barrier height and exothermicity for this reaction were, however, in poor agreement with experiment, which was attributed largely to deficiencies in the basis. Later calculations[38] with polarization functions added to the basis were made only for collinear structures and the predicted barrier height of 1.76 kcal mol^{-1} and exothermicity of 34.4 kcal mol^{-1} are in good agreement with experimental values (1.6 and 31.7 kcal mol^{-1}, respectively[39]).

The calculations of Bender *et al.*[38] (the so-called BOPS surface) have been used by several groups to provide potential energy functions for the reaction (6.13). Polanyi and Schreiber[39] fitted a LEPS function by a least-squares procedure to the 232 *ab initio* points for collinear HHF structures. The parameters were eight in total and occurred in the triplet-state potentials of both H_2 and HF[40]. The r.m.s. error was less than 1 kcal mol^{-1} and, apart from a shallow (spurious) well in the entrance valley, the fit was excellent.

Connor *et al.*[41] made small modifications to the BOPS surface to reproduce the experimental exothermicity and force constants of H_2 and HF. They then fitted the modified *ab initio* points for the collinear HHF structure to a rotated Morse function:

$$V(p, \phi) = D\{\exp[-2\beta(p-p_0)-2\exp(-\beta(p-p_0)]\} \quad 6.14$$

where D, β and p_0 are functions of the angle which the vector $\boldsymbol{p}$ makes in a skew coordinate system which diagonalizes the kinetic energy (see Figure 4.2). The origin of $\boldsymbol{p}$ is in the plateau region of the potential. A grid of 25 values of ϕ was used to provide a cubic spline interpolation of the parameters D, β and p_0 (see Section 3.1). This potential was used to calculate the quantum cross-sections for the reaction (6.13) in a linear mode, and comparison was made with similar calculations on empirical LEPS surfaces (similarly Morse-spline fitted) which had earlier been produced by Muckerman[42].

The LEPS function, even in its various modifications, has generally few adjustable parameters; this is one of its attractive features, but one should not expect too much from it. It has been known from several studies that if a LEPS function is devised to reproduce a saddle point structure A–B–C (say), then these same parameters may give a very poor representation of other transition states corresponding to structures A–C–B and B–A–C. This is the case for the H_2F system. The Muckerman V LEPS surface, which is thought to give the best representation of the dynamics of reaction (6.13)[14], is inadequate for the reaction

$$H + HF \rightarrow HF + H \quad 6.15$$

Ab initio calculations[43,44] supported by experiment[45] show that this reaction proceeds by a symmetric transition state of C_{2v} symmetry (bond angle *ca.* 105°) with an activation energy in the range 40–50 kcal mol^{-1}. The Muckerman V surface has a collinear transition state with an activation energy of only 1.2 kcal mol^{-1} (ref. 45), and other LEPS surfaces[39] give a similar low value[46]. Similar discrepancies between the *ab initio* and LEPS surface have been found for the H_2Cl system[47].

Carter and Murrell[48] used the BOPS potential[38] in conjunction with calculations on the C_{2v} structure by Wadt and Winter[43] to obtain an analytical potential for the whole H_2F surface using the many-body expansion functions described in Section 3.5. The first difficulty which has to be overcome is to make corrections to different sets of data to bring them into a similar level of consistency. In the first place the BOPS calculations on linear HHF structures[38] are of better quality than those from the same workers on non-linear structures[37]. However, in the expectation (hope) that the change in energy on bending an HHF structure is not very different for the two bases, the BOPS data can be extrapolated to non-linear structures. There is unfortunately no overlap between the configurations studied by Wadt and Winter and BOPS, but, as they use similar quality bases and both use wave functions which reproduce the major part of the correlation energy, it can be assumed that the two sets of data can be amalgamated without corrections.

By examining the energies and geometries of the HHF and HFH saddle points as a function of the angle $F\hat{H}H$, Carter and Murrell[48] concluded that the H_2F surface was two-valued with an intersection between $^2A_1$ and $^2B_2$ surfaces somewhere in the C_{2v} space. There are some qualitative molecular orbital arguments to support this but no confirmation by *ab initio* calculations. On further study we have considered their conclusion to be unlikely, and, as we shall

Table 6.4 Parameters of the potential function of H_2F (ref. 49). The functions used are similar to those referred to in Table 6.1 except that symmetry coordinates are used for the range function

$$T=(1-\tanh \gamma_1 S_1/2)(1-\tanh \gamma_3 S_3/2)$$

$$S_1=\sqrt{\tfrac{1}{2}}\,(R_1+R_2-R_1^0-R_2^0);\quad S_3=R_3-R_3^0$$

*Two-body terms:**

Species	D_e/eV	R_e/Å	a_1/Å^{-1}	a_2/Å^{-2}	a_3/Å^{-3}
HF	6.123	0.9168	4.216	3.965	3.835
H_2	4.747	0.7414	3.952	4.030	3.537

Three-body term:

$C_1=C_2$/Å^{-1}	−0.1762	$C_{1111}=C_{2222}$/Å^{-4}	0.2333
C_3	0.2998	C_{3333}	0.1321
$C_{11}=C_{22}$/Å^{-2}	−0.1701	$C_{1113}=C_{2223}$	0.0942
C_{33}	0.1801	$C_{1133}=C_{2233}$	−0.6231
$C_{13}=C_{23}$	−0.4760	$C_{1333}=C_{2333}$	−0.0137
C_{12}	0.6013	$C_{1112}=C_{1222}$	−0.5876
$C_{111}=C_{222}$/Å^{-3}	0.4625	C_{1122}	0.6978
C_{333}	1.0204	$C_{1123}=C_{1223}$	−0.2660
$C_{113}=C_{223}$	−0.7231	C_{1233}	0.5137
$C_{133}=C_{233}$	−0.8845	V^0/eV	5.4512
$C_{112}=C_{122}$	−0.6037	γ_1/Å^{-1}	2.3753
C_{123}	0.8338	γ_3	2.1026
		$R_{HF}^0=R_1^0=R_2^0$/Å	1.5117
		$R_{HH}^0=R_3^0$	1.1042

* The two-body terms may differ slightly from one potential to another in this book (cf. Table 6.1). This is due to the fact that the potentials were produced at different times and they have not all been optimized with the latest (best) two-body terms.

show, this view is supported by the fact that the *ab initio* data can be fitted to a single-valued function[49].

The *ab initio* calculations show that the ground-state has $^2A'$ symmetry in C_s configuration which ascends to $^2\Sigma^+$ in $C_{\infty v}$ and $^2A_1$ in C_{2v}. The Wigner–Witmer rules (Section 2.4) allow dissociation from this surface to ground-state products

$$H_2F(^2A') \rightarrow \begin{matrix} H_2(X, {}^1\Sigma_g^+) + F(^2P) \\ HF(X, {}^1\Sigma^+) + H(^2S) \end{matrix} \qquad 6.16$$

There are other surfaces that originate from $F(^2P) + H_2(X, {}^1\Sigma_g^+)$, but they are probably not important for the dynamics of reaction (6.13) as they are higher in energy over most of configuration space. On the basis of reaction (6.16) we can write the many-body expression for the potential as

$$V_{HHF} = V_{HF}^{(2)}(R_1) + V_{HF}^{(2)}(R_2) + V_{HH}^{(2)}(R_3) + V_{HHF}^{(3)} \qquad 6.17$$

The parameters of the potential function which were obtained by a least-squares optimization to the *ab initio* data are given in Table 6.4. It was assumed that the data in ref. 38 were accurate in the $H_2 + F$ valley up to and including the saddle point but in the HF + H valley the points were smoothly extrapolated to give the experimental exothermicity of reaction (6.13). Contours of the potential which show the HHF and HFH transition states are given in Figures 6.7 and 6.8. The

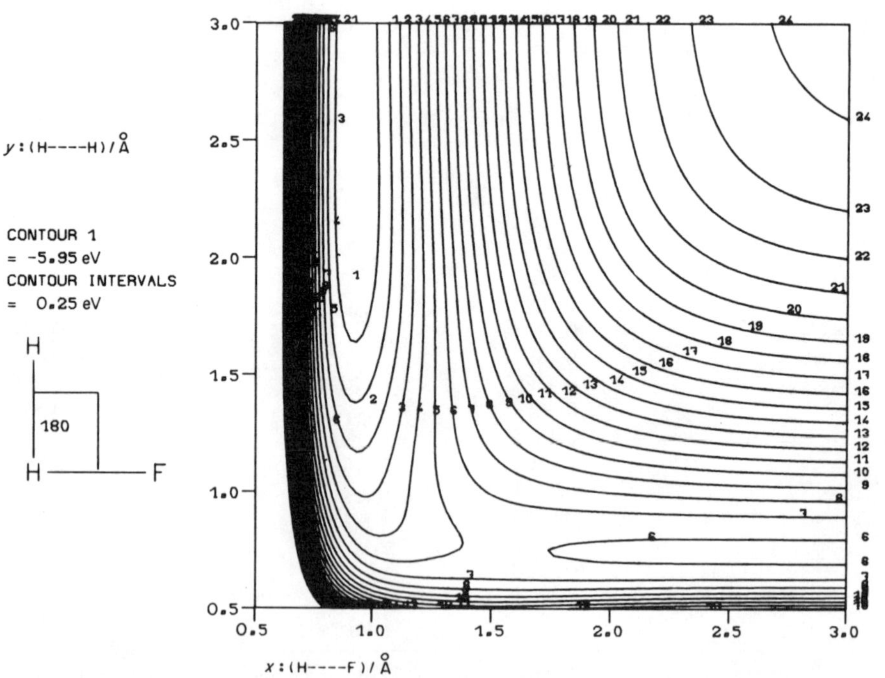

Figure 6.7 Contours of the HHF potential defined by the parameters of Table 6.3, showing the H–H–F transition state for reaction (6.13)

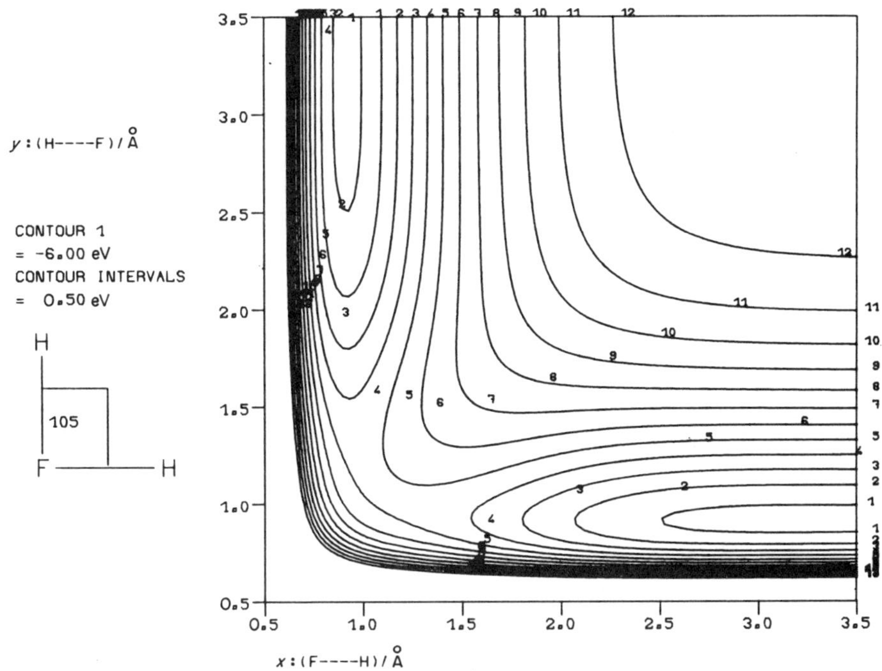

Figure 6.8 Contours of the HHF potential showing the transition state for the symmetric exchange reaction (6.15)

activation energy for reaction (6.15) obtained with this function is 1.66 kcal mol^{-1} and for the hydrogen exchange reaction 47 kcal mol^{-1}, both in excellent agreement with the best theoretical and experimental estimates.

This is an appropriate point to make some comparison between a LEPS function and the many-body expansion using Murrell–Sorbie functions (3.4). It is not surprising that the latter, which have more adjustable parameters, can provide a better overall fit to the potential surface. However, the balance of advantage must in a large part depend on how easily the functions can be used for dynamical calculations. Ease of use depends on two factors. Firstly, it is necessary to be able to differentiate the potential with respect to the variables used to solve the dynamical equations (e.g. to set up Hamilton's equation of motion). For both types of function this is a tedious process but we see little difference between the two. Secondly, one needs to evaluate the potential and its derivatives at general configuration points. This is usually the most time-consuming part of the classical trajectory calculations.

The Murrell–Sorbie functions require the evaluation of high-order polynomials and exponentials and will usually be more demanding of computer time than the LEPS functions which, with a Morse representation of the diatomics, only require a few exponentials to be evaluated. However, the Murrell–Sorbie functions have been used for many classical trajectory studies and they are not exceptionally expensive. Given their greater accuracy and flexibility, we feel that for most systems they are greatly to be preferred to LEPS functions.

References

1. E. H. Taylor and S. Datz, *J. Chem. Phys.*, **23**, 1711 (1955).
2. J. L. Kinsey, in *MTP International Review of Science*, Ed. J. C. Polanyi, Butterworths, London 1972, Vol. 9, p. 193; D. R. Herschbach, *Faraday Discuss. Chem. Soc.*, **62**, 162 (1977).
3. W. A. Lester and M. Krauss, *J. Chem. Phys.*, **52**, 4775 (1970).
4. D. R. Herschbach, in *Proceedings of Conference on Potential Energy Surfaces*, Ed. W. A. Lester, IBM Res. Lab., 1971, p. 44.
5. G. G. Balint-Kurti and R. N. Yardley, *Faraday Discuss. Chem. Soc.*, **62**, 77 (1977).
6. Y. Zeiri and M. Shapiro, *Chem. Phys.*, **31**, 217 (1978).
7. J. C. Polanyi, *J. Chem. Phys.*, **31**, 1338 (1959); B. A. Hudson and J. C. Polanyi, *J. Chem. Phys.*, **55**, 4745 (1971).
8. M. M. L. Chen and H. F. Schaefer, *J. Chem. Phys.*, **72**, 4376 (1980).
9. C. H. Becker, P. Casavecchio, P. W. Tiedemann, J. J. Valentini and Y. T. Lee, *J. Chem. Phys.*, **73**, 2833 (1980).
10. S. Carter and J. N. Murrell, *Mol. Phys.*, **41**, 567 (1980).
11. J. M. Alvarino, O. Gervasi and A. Lagana, *Chem. Phys. Lett.*, **87**, 254 (1982).
12. J. M. Alvarino, P. Casavecchio, O. Gervasi and A. Lagana, *J. Chem. Phys.*, **77**, 6341 (1982).
13. M. Schapiro and Y. Zeiri, *J. Chem. Phys.*, **70**, 5264 (1979).
14. R. B. Walker, Y. Zeiri and M. Shapiro, *J. Chem. Phys.*, **74**, 1763 (1979).
15. C. H. Wu, *J. Chem. Phys.*, **65**, 3181 (1976).
16. P. J. Foster, R. E. Leckenby and E. R. Robbins, *J. Phys. B*, **2**, 478 (1969); A. Herrmann, S. Leutwyler, E. Schumacher and L. Waste, *Helv. Chim. Acta*, **61**, 453 (1978).
17. D. M. Lindsay, D. R. Herschbach and L. A. Kwiram, *Mol. Phys.*, **32**, 1199 (1976).
18. R. Huber and H. G. Weber, *Chem. Phys.*, **37**, 173 (1979).
19. R. Gupta, W. Happer, G. Moe and W. Park, *Phys. Rev. Lett.*, **32**, 574 (1974).
20. H. G. Weber, H. J. Glas, R. Huber, M. Kompitsas, G. Schmidt and G. zu Putlitz, *Z. Phys.*, **268**, 91 (1974).
21. J. C. Whitehead and R. Grice, *Faraday Discuss. Chem. Soc.*, **62**, 255 (1977).
22. R. N. Yardley and G. G. Balint-Kurti, *Chem. Phys.*, **16**, 287 (1976).
23. W. H. Kendrick and I. H. Hillier, *Mol. Phys.*, **33**, 635 (1977).
24. P. S. Bagus, G. del Conde and D. W. Davies, *Faraday Discuss. Chem. Soc.*, **62**, 321 (1977); D. W. Davies and G. del Conde, *Chem. Phys.*, **12**, 45 (1976).
25. W. H. Gerber and E. Schumacher, *J. Chem. Phys.*, **69**, 1692 (1978).
26. J. L. Gole, R. H. Childs, D. A. Dixon and R. A. Eades, *J. Chem. Phys.*, **72**, 6368 (1980).
27. R. L. Martin and E. R. Davidson, *Mol. Phys.*, **35**, 1713 (1978).
28. A. J. C. Varandas, J. Tennyson and J. N. Murrell, *Chem. Phys. Lett.*, **61**, 431 (1979).
29. S. N. Mohammad, *Physica*, **C95**, 418 (1978); *Int. J. Quant. Chem.*, **13**, 429 (1978); *Z. Naturforsch., Teil A*, **33**, 18 (1977).
30. A. L. Companion, *Chem. Phys. Lett.*, **56**, 500 (1978).
31. A. J. C. Varandas and V. M. F. Morais, *Mol. Phys.*, **47**, 1241 (1982).
32. A. J. C. Varandas, *J. Chem. Soc., Faraday Trans. 2*, **76**, 129 (1980).
33. A. J. C. Varandas and J. Brandão, *Mol. Phys.*, **45**, 857 (1982).
34. J. C. Polanyi and J. L. Schreiber, 'Kinetics of Gas Reactions', in *Physical Chemistry, an Advanced Treatise*, Eds. H. Eyring, D. Henderson and W. Jost, Academic Press, New York, 1974, Vol. 6A.
35. M. J. Berry, *J. Chem. Phys.*, **59**, 6229 (1973).

36. T. P. Schafer, P. E. Siska, J. M. Parson, F. P. Tully, Y. C. Wong and Y. T. Lee, *J. Chem. Phys.*, **53**, 3385 (1970).
37. C. F. Bender, P. K. Pearson, S. V. O'Neil and H. F. Schaefer, *J. Chem. Phys.*, **56**, 4626 (1972).
38. C. F. Bender, S. V. O'Neil, P. K. Pearson and H. F. Schaefer, *Science*, **176**, 1412 (1974).
39. J. C. Polanyi and J. L. Schreiber, *Chem. Phys. Lett.*, **29**, 319 (1974).
40. L. Pederson and R. N. Porter, *J. Chem. Phys.*, **47**, 4751 (1967).
41. J. N. L. Connor, W. Jakubetz and J. Manz, *Mol. Phys.*, **29**, 347 (1975); **35**, 1301 (1978).
42. J. T. Muckerman, *J. Chem. Phys.*, **54**, 1155 (1971); **56**, 2997 (1972); personal communication, quoted in refs. 39 and 41.
43. W. R. Wadt and N. W. Winter, *J. Chem. Phys.*, **67**, 3068 (1977).
44. C. F. Bender, B. J. Garrison and H. F. Schaefer, *J. Chem. Phys.*, **62**, 1188 (1975).
45. F. E. Bartoszek, D. M. Manos and J. C. Polanyi, *J. Chem. Phys.*, **69**, 933 (1978).
46. J. N. L. Connor, *Comput. Phys. Commun.*, **17**, 117 (1979).
47. T. H. Dunning, *J. Chem. Phys.*, **66**, 2752 (1977).
48. S. Carter and J. N. Murrell, *Mol. Phys.*, **41**, 567 (1980).
49. S. Carter, unpublished work.

Chapter 7

Fitting to spectroscopic data

7.1 Fitting to force constants

In Section 1.2 we showed that good diatomic potentials could be obtained by taking a simple function and parameterizing it to reproduce the force constants, the equilibrium bond length and the dissociation energy of the molecule. This procedure will work fairly well with several types of function, but the one that we have found to be particularly suitable is the Extended Rydberg function (1.19). One important point to note is that, although the best potentials are obtained by fitting the quadratic, cubic and quartic force constants, reasonably satisfactory functions of the Rydberg type (1.17) are given by fitting the quadratic force constant alone. This is significant for the polyatomic problem because complete force fields to quartic terms are known for relatively few polyatomic molecules and it is more common that only the quadratic force constants are known.

The determination of anharmonic force fields for polyatomic molecules is a much more difficult task than for diatomic molecules and it was not until 1958 that the first analyses were made by Pliva[1]. In succeeding years several research groups have tackled the problem using the technique of perturbation theory and their work has been reviewed by Mills[2]. Although the most accurate analysis is made from high-resolution spectroscopic data, *ab initio* calculations can also give useful information on force fields.

We follow Mills and define the force constants, f, in terms of the multi-dimensional Taylor expansion about a potential minimum

$$\Delta V = \frac{1}{2!} f_{ij} r_i r_j + \frac{1}{3!} f_{ijk} r_i r_j r_k + \frac{1}{4!} f_{ijkl} r_i r_j r_k r_l + \cdots \qquad 7.1$$

where an independent summation is implied over each index (i.e. both terms f_{12} and f_{21} appear in the expansion). The reader must be vigilant in reading the literature because in some cases the factors $(1/n!)$ or $(2/n!)$ are subsumed into the force constant and sometimes the multiple summations are restricted ($j \leqslant i$, etc.). With the definition implied by equation (7.1) the force constants are equal to the derivatives of V at equilibrium:

$$f_{ijk\ldots} = (\partial^n V / \partial r_i \, \partial r_j \, \partial r_k \ldots) \qquad 7.2$$

Force fields are defined with respect to internal coordinates and these are usually a mixture of bond-stretching and angle-bending coordinates. Customary usage by experimentalists is to take bond-stretching coordinates in Å and angle-bending coordinates either in radians (dimensionless) or in Å by multiplying by an appropriate scaling factor which is usually an equilibrium bond length. The stretch–stretch quadratic force constant is normally expressed in the units mdyn Å^{-1} or, equivalently, aJ Å^{-2}.* Quantum chemists will often report their calculations in atomic units.

In Section 1.2 we showed that the force constants of diatomic molecules could be obtained from the spectroscopic parameters, ω_e, $\omega_e x_e$, B_e, α_e, etc., which relate the vibration–rotation energy levels to a power series expansion in the quantum numbers v and J as in expression (1.10). The analysis for polyatomic molecules follows the same route but is naturally more complicated because there is more than one quantum number for both the vibrational and rotational states of the molecule. In addition, there are two features which are absent from diatomic systems. The first is Coriolis coupling, which is described by a Hamiltonian containing a vector product of rotation and vibration angular momentum operators. This leads to various types of vibration–rotation interaction with a consequent effect on the spectroscopic parameters. The second is Fermi resonance, which is the coupling of different vibrational modes by anharmonic terms in the potential.

It is outside the scope of this book to describe the details of the perturbation approach to the spectroscopic parameters of polyatomic molecules and the reader is referred elsewhere for details[2]. We do, however, make the point that it is a difficult problem, both analytically and computationally; not least there is the uncertainty about the convergence of the perturbation expansion. We shall later in this chapter examine a variational approach to vibration–rotation frequencies which may be an easier and more accurate route to the determination of molecular force fields.

We now describe the procedure used by Sorbie and Murrell[3] to determine a potential function for a stable triatomic molecule by parameterizing the three-body term to the force field. We start with the simplest system, which has only a single minimum on the surface although, as we shall see, these are relatively rare.

If we write the total potential as a sum of one-, two and three-body terms as in equation (3.2) then, provided that the first two are known, the three-body term can be deduced from the difference between the total potential and the one- and two-body terms:

$$V^{(3)}_{\mathrm{ABC}}(R_1, R_2, R_3) = V_{\mathrm{ABC}}(R_1, R_2, R_3) - V^{(1)}_{\mathrm{A}} - V^{(1)}_{\mathrm{B}} - V^{(1)}_{\mathrm{C}} - V^{(2)}_{\mathrm{AB}}(R_1) - V^{(2)}_{\mathrm{BC}}(R_2) - V^{(2)}_{\mathrm{CA}}(R_3) \quad 7.3$$

The three-body term is written as a polynomial $P(R_1, R_2, R_3)$ in the three internuclear distances multiplied by a range function T of the form (3.26) or (3.27). Substituting in equation (7.3) and multiplying by T^{-1} gives

* 1 mdyn = 10^{-3} dyn = 10^{-8} N; 1 aJ (attojoule) = 10^{-18} J ≈ 6.2415 eV.

$$P(R_1, R_2, R_3) = V^{(3)}_{\text{ABC}}(R_1, R_2, R_3)T^{-1}$$
$$= [V_{\text{ABC}}(R_1, R_2, R_3) - V^{(1)}_{\text{A}} - V^{(1)}_{\text{B}} - V^{(1)}_{\text{C}}$$
$$- V^{(2)}_{\text{AB}}(R_1) - V^{(2)}_{\text{BC}}(R_2) - V^{(2)}_{\text{CA}}(R_3)]T^{-1} \quad 7.4$$

If the parameters γ_i in the range function are specified, then one can determine the coefficients of the polynomial from known properties of the ABC surface. If the functions chosen for P and T have sufficient flexibility to provide an exact representation of $V^{(3)}_{\text{ABC}}$, then the coefficients of P and the parameters γ_i can be determined from any set of ABC surface data, provided it is sufficiently numerous. As an exact representation of $V^{(3)}_{\text{ABC}}$ cannot be obtained from simple functions in practice, we have to make some decision on how equation (7.4) is to be used to obtain the optimum functions P and T within their constrained forms. The approach used by Sorbie and Murrell[3] was to make a Taylor expansion of the right-hand side of equation (7.4) about the equilibrium position and to write the polynomial P in displacement coordinates from equilibrium, ρ_i. The coefficients of P can then be obtained by equating terms of like power on both sides of equation (7.4).

We examine the details of this procedure by writing the polynomial in the form (6.5):

$$P = V^0\left(1 + \sum_i C_i\rho_i + \sum_i \sum_{j \leqslant i} C_{ij}\rho_i\rho_j + \cdots\right) \quad 7.5$$

and taking the range function as (3.26):

$$T = \prod_{i=1,3} (1 - \tanh \gamma_i\rho_i/2) \quad 7.6$$

where $\rho_i = R_i - R_i^0$, R_i^0 being the coordinates of the equilibrium configuration of the triatomic molecule. Noting that $T=1$ at $\rho_i=0$, we see immediately that

$$V^0 = V^{(3)}(R_1^0, R_2^0, R_3^0)$$
$$= V_{\text{ABC}}(R_1^0, R_2^0, R_3^0) - V^{(1)}_{\text{A}} - V^{(1)}_{\text{B}} - V^{(1)}_{\text{C}} - V^{(2)}_{\text{AB}}(R_1^0) - V^{(2)}_{\text{BC}}(R_2^0) - V^{(2)}_{\text{CA}}(R_3^0) \quad 7.7$$

We now consider, for example, the first derivative of equation (7.4) with respect to R_1 evaluated at the equilibrium configuration of the triatomic molecule; the result is

$$V^0 C_1 = \left(\frac{\partial V^{(3)}_{\text{ABC}}}{\partial R_1}\right)_0 + V^0\left(\frac{\partial T^{-1}}{\partial R_1}\right)_0 \quad 7.8$$

Noting that the first derivative of V_{ABC} at equilibrium is zero, we have, after differentiating T^{-1}, the result

$$V^0 C_1 = -\left(\frac{\mathrm{d}V^{(2)}_{\text{AB}}}{\mathrm{d}R_1}\right)_0 + \frac{V^0\gamma_i}{2} \quad 7.9$$

the derivative of the diatomic potential being evaluated at the equilibrium configuration of the triatomic *not* the diatomic species.

It is clear that by further differentiation we can set up a relationship between the coefficients of P and derivatives of $V^{(3)}_{\mathrm{ABC}}$, and Appendix 3 gives the appropriate expressions for the coefficients up to quartic terms. The partial derivatives of $V^{(3)}$ are, from equation (7.3), related to the derivatives of the two-body terms and to partial derivatives of the full triatomic potential, and the latter can be obtained from the force constants of the triatomic molecule. As force constants are normally expressed in internal coordinates, these must be converted to derivatives in (R_1, R_2, R_3) to obtain the necessary partial derivatives of $V^{(3)}$. The relevant expressions are given in Appendix 1.

The set of equations of which equations (7.7) and (7.9) are two members are linear in the coefficients C and can be solved by matrix inversion. However, this leaves the problem of determining the non-linear parameters γ_i. In Chapter 6 we described how these parameters could be optimized by a non-linear least-squares fitting to *ab initio* points on the surface, and if such points are available this is the procedure we would recommend even if the linear coefficients are to be determined from the spectroscopic force field. If *ab initio* points are not available then one must resort to some qualitative criteria for the optimization.

By choosing the coefficients of the three-body polynomial to reproduce the triatomic force field we obtain a potential which is correct in the region of the equilibrium configuration of the triatomic and is asymptotically correct at the dissociation limits. Whether it is correct in the regions between is a matter for examination. The chances of it being correct will probably be better if the complete force field to quartic terms is reproduced than if only the harmonic force field is reproduced, i.e. if a quartic polynomial is used in the three-body term rather than a quadratic polynomial. However, there are dangers in this assumption because the higher the order of the polynomial the more likely is it that there will be further stationary points in the potential between R_i^e and infinity and the most significant factor determining the existence of such stationary points is the value of the parameters γ_i.

The most important criterion that we can therefore impose for the γ_i is that they give no unphysical behaviour to the surface, in particular that there are no stationary points on the surface that are not supported by physical evidence. We shall see in later examples that there are many surfaces which have more than one minimum and there are often maxima along a bond-breaking reaction coordinate, so the existence of further points is by no means ruled out. Whether they are present or not is an important qualitative, in some cases quantitative, criterion for determining the γ_i.

7.2 A potential function for HO_2 ($\tilde{X}$, $^2A''$)

We have already mentioned the fact that there are few triatomic systems which have only one minimum on their ground state surface, and the few which do, such as H_2O, usually have other complications arising from surface intersections. One of the simplest examples we can give of a potential function which has been derived by fitting to force constants, according to the method of the previous section, is for the ground state of HO_2. The equilibrium configuration of this species has C_s symmetry, with an HOO bond angle of 106°, but by virtue of permutation symmetry there are two equivalent minima on the surface which differ only in respect of which oxygen atom has the hydrogen attached to it. To introduce this symmetry into the potential function we therefore use symmetry coordinates as variables as described in Section 4.2.

The ground state of HO_2 is $^2A''$ in the C_s group and is therefore antisymmetric to reflection in the molecular plane. For linear configurations this state correlates with one component of a $^2\Pi$ state but we do not in this section deal with the other component which is a $^2A'$ excited state of the bent molecule. The $^2A''$ state dissociates to ground-state products according to the following scheme,

$$\mathrm{HO_2}(\tilde{X}, {}^2A'') \rightarrow \begin{cases} \mathrm{O_2}(X, {}^3\Sigma_g^-) + \mathrm{H}(^2\mathrm{S}) \\ \mathrm{OH}(X, {}^2\Pi) + \mathrm{O}(^3\mathrm{P}) \\ \mathrm{H}(^2\mathrm{S}) + 2\mathrm{O}(^3\mathrm{P}) \end{cases} \qquad 7.10$$

hence the surface can be represented by a single-valued function for all configurations.

A harmonic force field for HO_2 has been determined by Ogilvie[4] by analysing the infrared frequencies of the species (including several isotopes) trapped in a low-temperature matrix. A further refinement of the data has been made by Carter and Mills[5,6]. The six harmonic force constants, the dissociation energy[7] and the three parameters of the equilibrium geometry provide ten items of data from which the coefficients of the three-body polynomial can be determined; these data are summarized in Table 7.1.

Table 7.1 Geometry and force field of HO_2 (refs. 4–7).

R_1(OH)/Å	0.985	f_{11}/aJ Å^{-2}	6.502	$f_{\alpha\alpha}$/aJ	1.048
R_2(OO)	1.360	f_{22}	5.839	$f_{1\alpha}$/aJ Å^{-1}	−0.5
α/°	106.0	f_{12}	0.1	$f_{2\alpha}$	0.397
D_e/eV	7.474				

To use symmetry coordinates the reference structure from which the displacement coordinates $\rho_i = R_i - R_i^0$ are defined must have C_{2v} symmetry, and the R_i^0 are most conveniently taken as the average of the bond lengths in the two equivalent equilibrium configurations:

$$\left.\begin{aligned} R_1^0 &= R_3^0 = \tfrac{1}{2}(0.985 + 1.886) = 1.436\ \text{Å} \\ R_2^0 &= 1.36\ \text{Å} \end{aligned}\right\} \qquad 7.11$$

The symmetry coordinates are then

$$S_1 = \sqrt{\tfrac{1}{2}}(\rho_1 + \rho_3)$$

$$S_2 = \sqrt{\tfrac{1}{2}}(\rho_1 - \rho_3)$$

$$S_3 = \rho_2 \qquad 7.12$$

and in these coordinates the two equivalent C_s minima on the surface are at $(0, -a, 0)$ and $(0, a, 0)$ where $a = 0.637$ Å. The relationship between the two minima and the reference structure is illustrated in Figure 7.1.

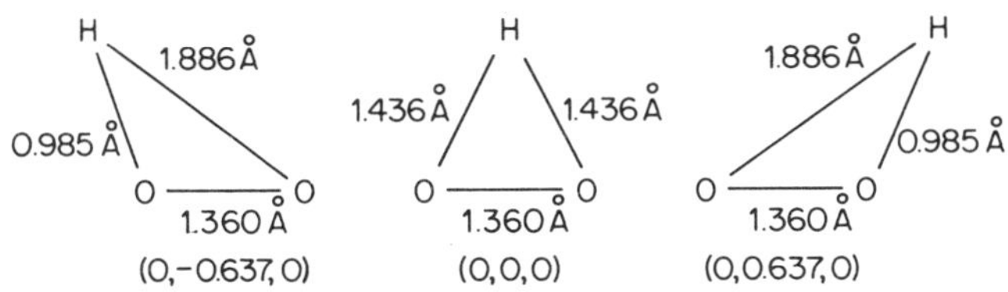

Figure 7.1 The two equivalent equilibrium structures of HO_2 and (centre) the reference structure of C_{2v} symmetry for the many-body expansion. The S_i coordinates of each structure are shown under the structures

As explained in Section 4.3 and summarized in Appendix 2 (Table A2.3), the data in Table 7.1 are sufficient to determine the coefficients of the following terms in the three-body polynomial: the constant, the two linear symmetric terms S_1 and S_3, the four symmetric quadratic terms S_1^2, S_2^2, S_3^2 and S_1S_3 and three higher-order terms $S_1S_2^2$, $S_3S_2^2$ and S_2^4. The range function must also involve only the symmetric coordinates and is given the form (3.26):

$$T = [1 - \tanh(\gamma_1 S_1/2)][1 - \tanh(\gamma_3 S_3/2)] \qquad 7.13$$

The relationship between the coefficients of the polynomial and the parameters of Table 7.1 [cf. equations (7.7) and (7.9)] is only slightly complicated by the fact that the equilibrium structures to which the data refer do not coincide with the reference structure. As T depends only on the variables S_1 and S_3, both of which are zero at the equilibrium structure, the derivatives of T^{-1} at $S_1 = S_2 = 0$ depend only on the values of γ_1 and γ_3, and not on the equilibrium geometry.

Surfaces were examined for a range of values of γ_1 and γ_3 and the values $\gamma_1 = 1.5$ Å and $\gamma_3 = 1.0$ Å were judged to give the best surface when all configurations were examined. However, this conclusion must be viewed in the knowledge that there have been no *ab initio* studies of the HO_2 surface away from the minima and in the absence of such studies we have no data on such important features of the surface as the transition state between the two C_s minima.

The parameters of the three-body term for the HO_2 potential are given in Table 7.2, and Figures 7.2 and 7.3 show contours of the potential for H approaching O_2 and O approaching OH, all species being in their ground states. There are no barriers to formation of the equilibrium structure for a wide range of approach

Table 7.2 Potential energy function for HO_2. The function has the same form as that described in Table 6.1 except that the three-body variables are S_i (7.12) rather than ρ_i.

Two-body terms:

Species	D_e/eV	R_e/Å	a_1/Å^{-1}	a_2/Å^{-2}	a_3/Å^{-3}
O_2	5.2132	1.2075	6.080	11.477	11.003
OH	4.6211	0.9696	4.507	4.884	3.795

Three-body term:

C_1/Å^{-1}	0.2496	C_{122}/Å^{-3}	0.2166
C_3	0.2023	C_{223}	−2.4219
C_{11}/Å^{-2}	0.7606	C_{2222}/Å^{-4}	2.3482
C_{22}	−1.9352	V^0/eV	4.4393
C_{33}	0.7291	γ_1/Å^{-1}	3.0
C_{13}	−1.2410	γ_3	2.0
		R^0_{OH}/Å	1.436
		R^0_{OO}	1.36

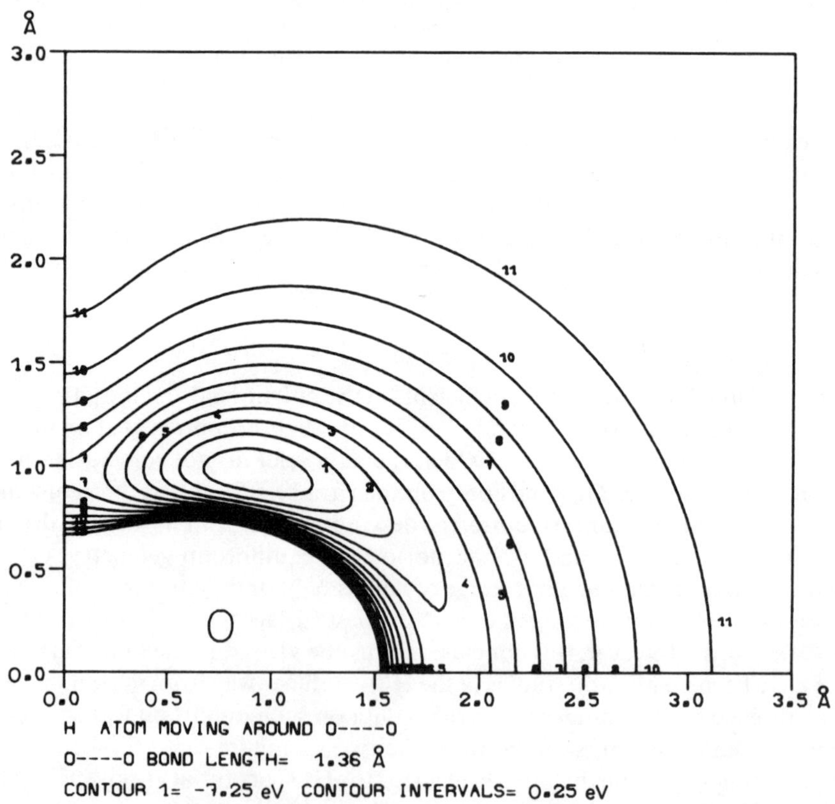

Figure 7.2 Contours of the HO_2 potential given in Table 7.2, for H moving around O_2. Only one quarter of the projection on to a plane passing through the three nuclei is shown, as the complete plane is obtained by symmetry

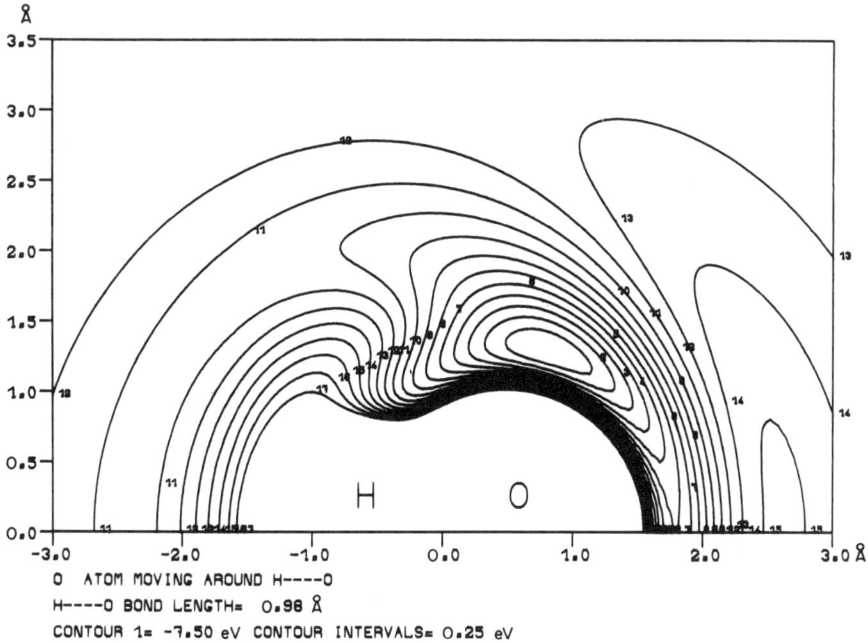

Figure 7.3 Contours of the HO_2 potential for O moving around HO

angles in both cases, and this is in accord with the fact that radical–radical reactions are generally very fast.

Quasi-classical trajectories have been calculated[8] for the reaction

$$H + O_2(^3\Sigma_g^-) \rightarrow OH(^2\Pi) + O(^3P) \qquad 7.14$$

on an earlier version[6] of the HO_2 surface than that given in Table 7.2, but one not appreciably different in form. This reaction is endothermic and the calculations predict the rate to be enhanced if the energy of the reactants is in vibrational or rotational modes of the O_2 rather than in its translational motion. The reaction has also been studied on a modified LEPS surface[9] which is less accurate in the region of the HO_2 equilibrium configuration, but without further molecular beam studies it will not be possible to make a detailed comparison of the dynamical predictions obtained from the two surfaces.

7.3 Fitting to frequencies; application to SO_2 ($\tilde{X}$, $^1A_1$)

There are two ways in which the vibrational energy levels of a molecule can be calculated if its potential energy function is known. The first is by perturbation theory: the first approximation to the frequencies is obtained by making a normal coordinate analysis on the basis of the harmonic approximation to the surface and changes in these frequencies due to anharmonicity are calculated using standard perturbation theory. It is usually by a trial-and-error study with this method that force constants are determined from observed frequencies. The second approach is to use the variational method in which the energy is minimized with respect to variation in a trial vibrational wave function. For success by this method one requires a basis from which a good trial function can be constructed and the harmonic approximation is again generally a good starting point.

We shall not give the details of the variational method in this book[10]. For triatomic molecules with rather rigid structures, either bent or linear, an appropriate vibration–rotation Hamiltonian is known and the basis sets are either harmonic oscillator eigenfunctions of the form

$$\Phi_n(Q) = H_n(Q) \exp(-Q^2/2) \qquad 7.15$$

where H_n is a Hermite polynomial and Q a normal coordinate, or, for the bending mode of a linear molecule, the basis functions involve associated Laguerre polynomials. For a floppy triatomic molecule, i.e. a molecule with a bent equilibrium configuration but a low barrier to linearity, the appropriate Hamiltonian and basis functions are more complicated[11,12]. In all cases one must ensure that sufficient basis functions are included in the variation method to obtain convergence.

In some of the first variational studies of vibrational frequencies[13], it was noted that the Taylor expansion of the potential about the equilibrium configuration in which the parameters (force constants) were determined by the perturbation method did not give very good vibration frequencies. At the harmonic level this was to be expected, but it was also true for some supposedly accurate quartic force fields. The reason for this is that the Taylor expansion is a poor representation of the potential except for very small displacements from equilibrium, as noted in Section 1.2 for diatomic molecules (see Figures 1.1 and 1.2). However, it was also noted that the H_2O potential deduced by Sorbie and Murrell[3] by fitting the quartic force field gave much better frequencies than the Taylor expansion using the same parameters[3]. The reason is that the Sorbie–Murrell potential has the correct asymptotic limits at dissociation and is therefore a much better representation of the complete potential surface. It was later shown that potentials of this general form fitted only to harmonic force constants gave respectable variational energies, and the implications of this are that the imposition of the correct asymptotic limits introduces approximately the correct anharmonicity into the surfaces.

No matter how carefully one constructs a potential according to the procedure described in the previous section, it is limited by the accuracy of the force field, which in turn is based on assumptions about the termination of the spectroscopic (Dunham) and perturbation expansions. For example, the HO_2 potential whose parameters are given in Table 7.2 gives variationally derived fundamental frequencies of 1090, 1230 and 3116 cm^{-1}, which should be compared with the experimental frequencies of 1101, 1389 and 3414 cm^{-1} (ref. 14) on which the harmonic force field was based. With this knowledge it is natural to think of optimizing the parameters of the potential directly to the variationally calculated vibration frequencies rather than to the force constants. The procedure is inevitably more time consuming than fitting to force constants because the variational calculation is expensive if carried out to convergence. A procedure has been suggested by Carter *et al.*[15] for economy which involves iteration of the parameters to an inexpensive, but non-converged, variational calculation (i.e. with a small basis) with appropriate corrections for the lack of convergence. The correction factors are evaluated infrequently (say, every tenth cycle) by a converged calculation. We now describe the application of such a procedure to the ground state of SO_2.

The ground state of SO_2 is a singlet for all nuclear configurations and satisfies the following dissociation limits:

$$SO_2(\tilde{X},\,^1A_1) \rightarrow \begin{matrix} SO(X,\,^3\Sigma^-) + O(^3P) \\ O_2(X,\,^3\Sigma_g^-) + S(^3P) \end{matrix} \rightarrow 2O(^3P) + S(^3P) \qquad 7.16$$

The most stable configuration of the molecule has C_{2v} symmetry, but there have been several proposals, based on kinetic and spectroscopic observations, for the transient existence of metastable isomers. The experimental evidence is only circumstantial but it has prompted *ab initio* studies of the potential surface[6,16,17] and these provide very strong evidence for the existence of metastable isomers. The most thorough work is that by Dunning and Raffenetti[16];* they used a large basis set with polarization functions for both the oxygen and sulphur atoms and configuration interaction from a generalized valence bond (GVB) reference state. They found one C_s (superoxide) isomer SOO, with a bond angle $S\hat{O}O \approx 120°$, and a ring structure of C_{2v} symmetry with bond angles approximately 60°. The two metastable forms are predicted to have similar energies, about 4 eV above that of the stable form but just below the SO + O dissociation limit of 5.73 eV.

A close analogy can be drawn between the potential surface of SO_2 and that of the isovalent species O_3 which will be discussed in detail in Section 10.1. The stable structure of O_3 has C_{2v} symmetry, but by virtue of the permutation symmetry of the surface there must be three equivalent minima on the surface of this form. The analogues on the SO_2 surface are the stable structure and the two equivalent SOO structures. There is also, as we shall see, strong evidence from calculations for the existence of a D_{3h} (ring) form of O_3 and the analogue of this is the ring form of SO_2 predicted by Dunning and Raffenetti[16].

The first *ab initio* calculations on SO_2 (ref. 6) were at the SCF level using a small basis, and the predicted geometry and energy of the SOO isomer are certainly inferior to the predictions of the most recent work (in particular the predicted energy was much too low). There was also no indication from these calculations for the ring structure. These early calculations have been used, in combination with the energy geometry and force constants of SO_2, to produce a potential function using the technique described in the previous section. This potential was later refined by optimizing the variationally calculated to the experimental vibration frequencies of SO_2 (ref. 15). As this provides a good illustration of the

* These workers gave a good summary of the experimental evidence for the metastable species.

technique we wish to describe in this section, we shall examine the potential in some detail, although we recognize that the more recent *ab initio* calculations certainly imply that it is not accurate for large displacements from the SO_2 minimum.

The experimental dissociation energy, geometry and harmonic force field of SO_2 provide seven independent pieces of data on which a potential function can be based (the energy, two independent geometrical parameters and four independent force constants). Adding to this the calculated energy and geometry of the SOO isomer gives a total of eleven pieces of data, and these can be used to determine the coefficients of eleven terms in the three-body polynomial. The choice of polynomial terms is most conveniently made using the S_i variables which treat the three minima on the surface being considered in an even-handed way.

Table A2.3 in Appendix 2 gives the terms in the polynomial whose coefficients can be determined from the data, namely the constant, S_1, S_3, S_1^2, S_2^2, S_3^2, S_1S_2, S_3^3, $S_1S_3^2$, $S_2^2S_3$ and S_2^4. The S_i variables are determined by the geometries of the SO_2 and SOO isomers using the conditions of Table A2.1 [case (v)]. From the experimental geometry of SO_2 ($R_{SO}=1.431$ Å, $\hat{OSO}=119°$) and the calculated geometry of SOO (ref. 6) ($R_{SO}=1.69$ Å, $R_{OO}=1.27$ Å, $\hat{SOO}=126°$) we find

$$\begin{bmatrix} S_1 \\ S_2 \\ S_3 \end{bmatrix} = \begin{bmatrix} 0.534 & 0.655 & 0.534 \\ 0.707 & 0 & -0.707 \\ 0.463 & -0.755 & 0.463 \end{bmatrix} \begin{bmatrix} \rho_1 \\ \rho_2 \\ \rho_3 \end{bmatrix} \qquad 7.17$$

where $$\rho_1=\rho_3=R_{SO}-2.167 \text{ Å}; \quad \rho_2=R_{OO}-1.270 \text{ Å}$$

A potential was first obtained which was similar to that of HO_2 (Table 7.3) except for the addition of one more term in the three-body polynomial. It was found that a reasonably good potential could be obtained with a single non-linear variable in the range function, $\gamma_1=4.0$ Å^{-1} (the multiplier of S_1). By symmetry $\gamma_2=0$ (as for HO_2), but it was also found to be reasonable to take $\gamma_3=0$ also.

The potential was then optimized to the experimental vibration frequencies of SO_2 (ref. 18) (16 frequencies in all) whilst maintaining the same geometries and energies of the SO_2 and SOO structures[15]. This was done by adding further flexibility to the three-body term in two stages, firstly by adding additional polynomial terms to bring it up to a full quartic (a further 11 terms) and secondly by relaxing both the reference structure and rotation matrices by which the S_i variables are defined. The completely optimized potential is given in Table 7.3[15]. It should be stressed that the main emphasis in deriving this potential was to obtain good SO_2 vibration frequencies and, except for the requirement of a specified energy and geometry of SOO, there was no imposition of the behaviour in other regions of the surface. The success in fitting frequencies can be seen from Table 7.4, where the errors are no greater than 4 cm^{-1} even for the highest levels. The potential is therefore expected to give reliable predictions for those frequencies which have not yet been determined experimentally.

Table 7.3 Many-body potential for the ground state of SO_2 (ref. 15). As the three-body polynomial is a complete quartic, it can be expressed equivalently in ρ_i or S_i variables. In the table, ρ_i variables are used and permutation symmetry is assumed, i.e. $C_1 = C_3$, etc. The three-body range function, however, is expressed in S_i variables as in equation (7.13).

Two-body terms:

Species	D_e/eV	R_e/Å	a_1/Å^{-1}	a_2/Å^{-2}	a_3/Å^{-3}
SO	5.4296	1.4811	4.5180	5.4464	3.8670
O_2	5.2136	1.2074	5.4447	7.7828	5.9920

Three-body term:

C_1/Å^{-1}	2.3952	C_{1111}/Å^{-4}	−0.2873
C_2	−1.2112	C_{2222}	−0.1941
C_{11}/Å^{-2}	1.5418	C_{1112}	0.5587
C_{22}	−0.3415	C_{1122}	−0.5287
C_{12}	−1.5798	C_{1222}	0.2915
C_{13}	3.1987	C_{1113}	−0.2469
C_{111}/Å^{-3}	−0.1759	C_{1133}	−1.2033
C_{222}	−0.0608	C_{1123}	1.1119
C_{112}	0.2314	C_{1223}	−1.0322
C_{122}	−0.1245	V^0/eV	−3.0846
C_{113}	0.5586	γ_1/Å^{-1}	6.2179
C_{123}	−1.3218	γ_2	0.0
		γ_3	2.7655
		R_1^0(SO)/Å	2.2218
		R_2^0(OO)	2.8691

$S_1 = 0.3743\rho_1 + 0.8484\rho_2 + 0.3743\rho_3$
$S_2 = 0.7071\rho_1 - 0.7071\rho_3$
$S_3 = 0.5999\rho_1 - 0.5293\rho_2 + 0.5999\rho_3$

By differentiating the potential function of Table 7.3 and evaluating derivatives at the equilibrium geometry of SO_2, we can obtain a force field of the molecule. This is compared with the spectroscopically determined force fields of Strey and Mills[19] in Table 7.5. The agreement is very good, significant differences occurring mainly in the quartic terms.

Figure 7.4 shows contours of the potential for the bond stretching modes of SO_2 with the bond angle held at 119°. Figure 7.5 shows a more complete representation of the anharmonicity of the surface around the equilibrium configuration from the movement of O around SO with the diatomic bond length optimized between its value in SO_2 and its calculated value in SOO. Our qualification regarding the accuracy of this surface far from the SO_2 minimum must be remembered. The SOO minimum imposed on the potential can be seen from Figure 7.5, but there is a further minimum which is an artefact of the potential; it could be considered as a rather poor representation of the ring structure predicted by Dunning and Raffenetti[16].

Careful examination of the contours of the potential for the approach of S to

Table 7.4 Comparison of observed vibrational frequencies of SO_2 (cm^{-1}) (ref. 18) with the variationally calculated frequencies from the potential in Table 7.3. The levels are identified by the dominant component of the variational wave function (n_1, n_2, n_3), where n_1, n_2 and n_3 are the quantum numbers of symmetric stretch, bend and antisymmetric stretch, respectively.

Level	Calc.	Obs.	Level	Calc.	Obs.
100	1151	1151	012	3228	
010	515	517	111	3015	3011
001	1362	1362	400	4564	
200	2296	2296	040	2036	
020	1026		004	5385	
002	2714	2716	310	3950	
110	1667	1665	220	3324	
101	2499	2500	130	2685	
011	1877	1876	301	4754	4751
300	3433	3431	202	4955	
030	1533	1535	103	5166	5166
003	4054	4054	031	2893	
210	2812	2808	022	3737	
120	2178		013	4568	
201	3629	3630	211	4146	
102	3837		121	3525	
021	2387		112	4352	

Table 7.5 Comparison of the SO_2 force field obtained from a perturbation analysis of spectroscopic data by Strey and Mills[19], with derivatives of the potential of Table 7.3 at the equilibrium configuration[15]. Units are aJ Å^{-2} for f_{rr}, aJ for $f_{\alpha\alpha}$, etc. Values in parentheses were assumed zeros.

Parameter	Spectroscopic analysis	Many-body potential	Parameter	Spectroscopic analysis	Many-body potential
f_{rr}	10.420	10.410	f_{rrrr}	378.9	412.9
$f_{\alpha\alpha}$	1.680	1.640	$f_{\alpha\alpha\alpha\alpha}$	−15.4	−7.8
$f_{rr'}$	0.124	0.120	$f_{rrrr'}$	0.5	−2.6
$f_{r\alpha}$	0.525	0.563	$f_{rrr'r'}$	3.9	3.5
f_{rrr}	−71.2	−69.8	$f_{rrr\alpha}$	(0)	−1.9
$f_{\alpha\alpha\alpha}$	−2.4	−2.3	$f_{rr\alpha\alpha}$	9.0	−2.0
$f_{rrr'}$	−2.1	0.5	$f_{r\alpha\alpha\alpha}$	(0)	0.0
$f_{rr\alpha}$	−3.3	1.1	$f_{rrr'\alpha}$	(0)	1.4
$f_{r\alpha\alpha}$	−2.8	0.1	$f_{rr'\alpha\alpha}$	5.2	−0.5
$f_{rr'\alpha}$	−1.3	1.9			

O_2 shows that there is a barrier to the formation of equilibrium SO_2 which is 0.057 eV for C_{2v} approaches and 0.023 eV for $C_{\infty v}$ approaches. Classical trajectories on this surface for the reaction[20]

$$S(^3P) + O_2(^3\Sigma_g^-) \rightarrow SO(^3\Sigma^-) + O(^3P)$$

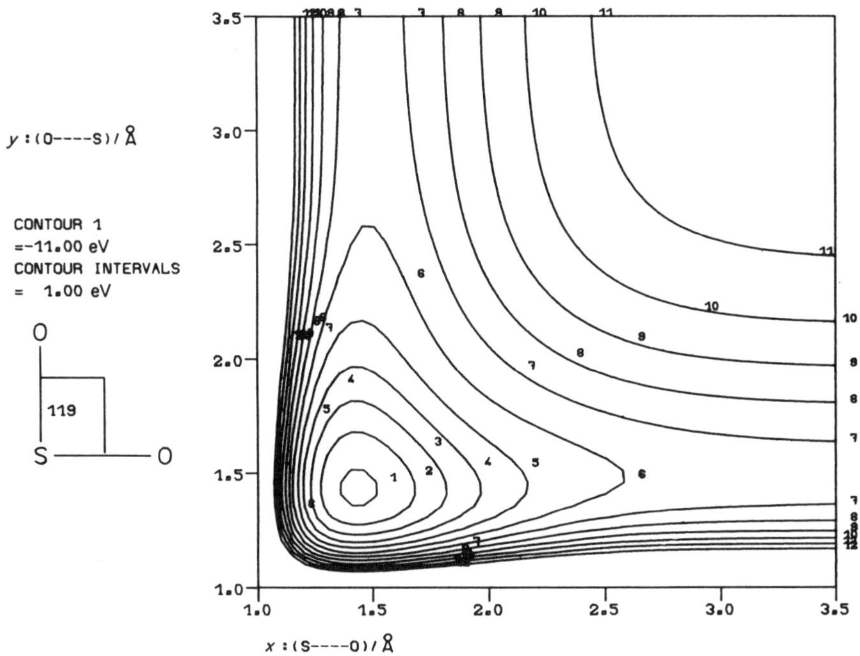

Figure 7.4 Contours of the SO_2 potential of Table 7.3 for a bond angle of 119°

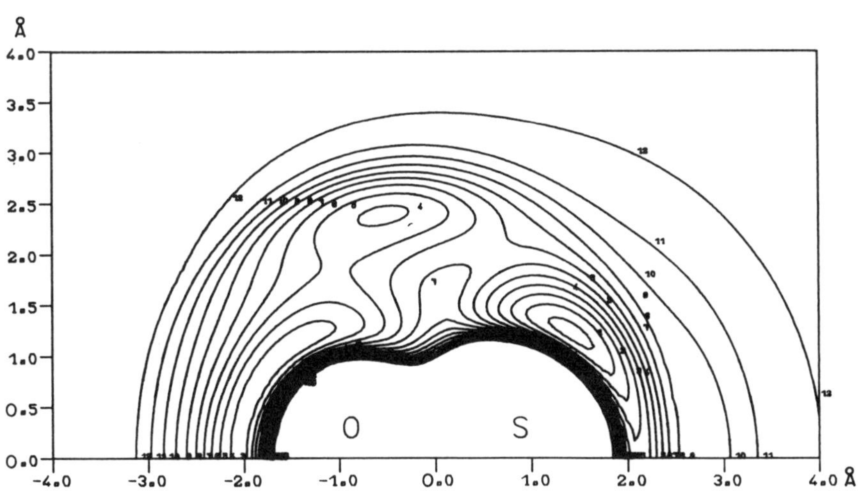

Figure 7.5 Contours of the SO_2 potential of Table 7.3 for O moving around SO with the diatomic bond length optimized for all positions of the atom

gave rate constants which were in reasonable agreement with the experimental values, but the predicted small activation energy was not found in experiments. It is probable that there should be no barrier for this reaction, at least for configurations which are close to collinear. Examination of the potential for O approaching SO shows no barrier in the exit valley of the reaction.

7.4 Linear molecules; application to C_3 and HCN

Bent triatomic molecules have three components of rotation and three modes of vibration. Linear triatomics have only two rotational components, two linear vibrational modes and a doubly degenerate bending vibrational mode. The analysis of the rotational–vibrational spectra of linear molecules is therefore different from that of bent molecules and the basis sets which are employed to describe the wave functions of these states are also different.

Appendix 1 gives the relationships between derivatives of the potential of a triatomic molecule in the coordinate system (R_1, R_2, R_3) and derivatives expressed in the spectroscopic coordinates (r_1, r_2, α). If these expressions are re-written to give derivatives of V in terms of those of W, then singularities will occur for linear molecules ($\alpha=\pi$) (ref. 3). For example:

$$\frac{\partial V}{\partial R_3}=\frac{\partial W}{\partial \alpha}\cdot\frac{\partial \alpha}{\partial R_3} \qquad 7.18$$

where

$$\frac{\partial \alpha}{\partial R_3}=-\frac{R_3}{R_1 R_2 \sin\alpha} \qquad 7.19$$

Thus equation (7.18) cannot be used directly to obtain $(\partial V/\partial R_3)$ for linear molecules; although $(\partial W/\partial \alpha)$ is zero, by symmetry, $(\partial \alpha/\partial R_3)_0$ is infinite. However, the expressions given in Appendix 1 are still valid for linear molecules and reduce to a simpler form which is given in Appendix 1a. It can be seen that an nth derivative of V with respect to R_3 is related to a 2nth derivative of W with respect to α.

7.4.1 The C_3 surface

The ground-state equilibrium structure of C_3 is linear ($D_{\infty h}$ point group) and is a singlet. As the ground state of C_2 is a singlet but that of the carbon atom is a triplet, the ground state of C_3 cannot dissociate adiabatically to ground-state species; either C_2 or C must be in an excited state. Possible dissociation channels which satisfy the spin conservation rules are

$$C_3(\tilde{X}, {}^1\Sigma_g^+)\rightarrow\begin{cases}C_2(a, {}^3\Pi_u)+C({}^3P) & \text{(a)}\\ C_2(X, {}^1\Sigma_g^+)+C({}^1D) & \text{(b)}\end{cases} \qquad 7.20$$

However, the triplet excited state of C_2 is only 0.089 eV above the singlet ground state at its equilibrium bond length[21], whereas the ${}^1D$–${}^3P$ separation of the carbon atom is 1.264 eV. It is therefore unlikely that there is any region of the surface where (b) is the lower energy channel. Accordingly, a potential energy function for C_3 can be produced in which the two-body terms are all taken from the $C_2(a, {}^3\Pi_u)$ state.

We note that the combination of a ${}^3\Pi_u$ and a ${}^3P$ state has a total multiplicity of $6 \times 9 = 54$ and only one of these connects to the C_3 ground state. Amongst the 54 states we have spin multiplicities of 1, 3 or 5 and orbital angular momenta of 0 (Σ), 1(Π) or 2 (Δ).

The equilibrium geometry and vibration fundamental bands of C_3 have been determined from gas-phase[22] and matrix isolation[23] spectroscopy, and a harmonic force field has been determined. However, the bending force constant is extremely low and the bending potential is very anharmonic. There have been two *ab initio* studies of the C_3 surface[24,25], displacements from equilibrium along the symmetric stretch, antisymmetric stretch and bend all having been examined. Although the data are not of very high accuracy they were used, together with the spectroscopic data, in a least-squares fitting procedure to determine the potential function[26].

The three-body term can be most simply constructed using the symmetry coordinates for an X_3 system (4.13) and totally symmetric combinations such as (4.23). The polynomial was complete to quartic terms. The range function was given the functional form (3.26). The potential is defined by the parameters in Table 7.6[26].

Figure 7.6 shows contours of the potential for moving one carbon atom around a C_2 molecule in its equilibrium configuration. The very low bending potential is evident from this figure and from Figure 7.7, which shows the bending potential for fixed C—C bond lengths.

The permutation symmetry of the C_3 surface is most clearly seen in a triangular graph with the coordinates shown in Figure 7.8. For a constant value

Table 7.6 Many-body potential for the ground state of C_3 (ref. 26). The three-body term has been written in ρ_i variables and permutation symmetry is implied ($C_1 = C_2 = C_3, \gamma_1 = \gamma_2 = \gamma_3$, etc.).

Two-body term:

Species	D_e/eV	R_e/Å	a_1/Å^{-1}	a_2/Å^{-2}	a_3/Å^{-3}
C_2	6.2413	1.3117	4.2009	4.0620	2.2526

Three-body term:

C_i/Å^{-1}	−5.5132	C_{iiii}/Å^{-4}	−4.5678
C_{ii}/Å^{-2}	−7.2255	C_{iiij}	12.5519
C_{ij}	−11.8935	C_{iijj}	9.7119
C_{iii}/Å^{-3}	9.8264	C_{iijk}	−21.6913
C_{iij}	−9.4647	V^0/eV	0.7296
C_{ijk}	−29.9286	γ_i/Å^{-1}	3.0
		R_i^0/Å	1.7025

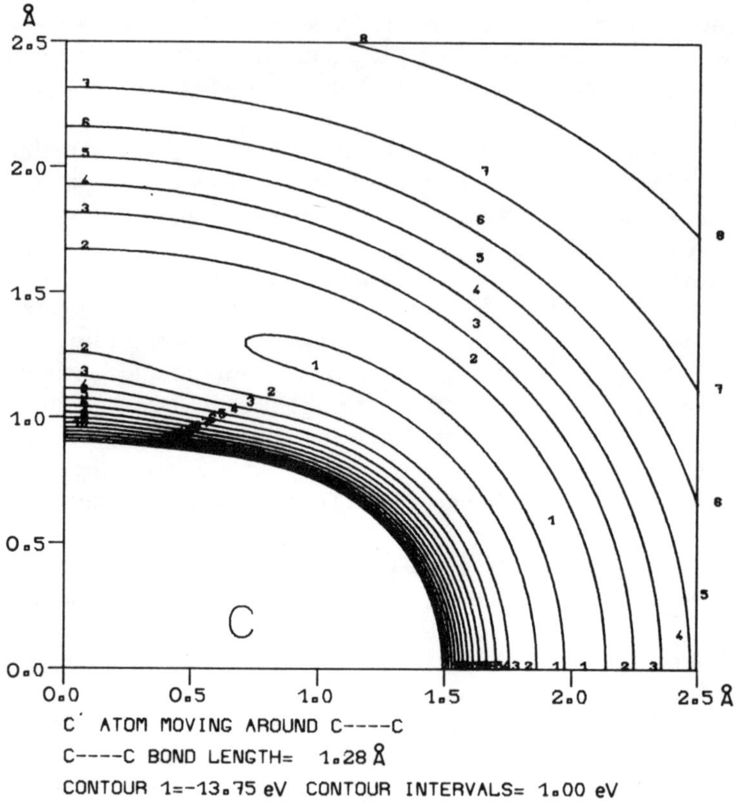

Figure 7.6 Contours of the C_3 potential given in Table 7.6 for one atom moving around the other two with a fixed bond length of 1.28 Å

of the perimeter of the X_3 triangle (S_1 = constant) such a graph will show the contours of V with variables S_2 and S_3. The space is bounded physically by the hatched area within which the triangle rule $R_i \geqslant R_j + R_k$ is satisfied. The boundaries of this space represent linear structures; C_{2v} structures lie on the lines joining apices to the centres of the opposite sides, and the origin ($S_2 = S_3 = 0$) represents a D_{3h} structure. Figure 7.9 shows such a graph for the C_3 surface with the value of S_1 equal to that for the equilibrium C_3 structure. The three equivalent equilibrium structures are represented by the minima centred at the edges of the triangular boundary. Note that the transition states for isomerization have C_{2v}, not D_{3h} structures.

7.4.2 *The HCN surface*

The potential energy surface of the ground state of HCN has been extensively studied by *ab initio* methods, the stimulation being the structure and stability of the metastable isocyanide HNC and the isomerization of this to the stable cyanide. Although alkyl cyanides are kinetically stable under normal conditions,

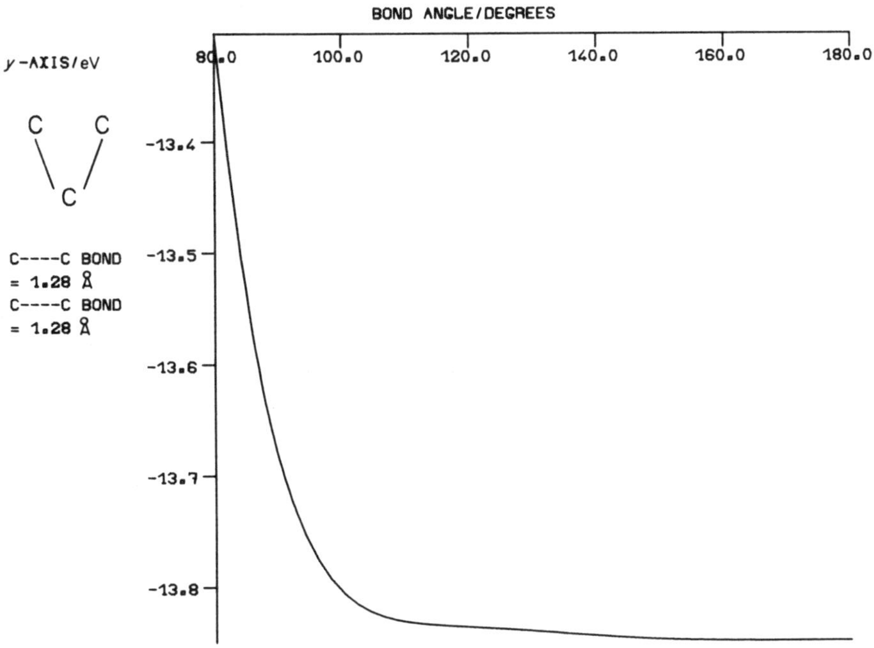

Figure 7.7 The bending potential for C_3 with fixed C—C bond lengths of 1.28 Å

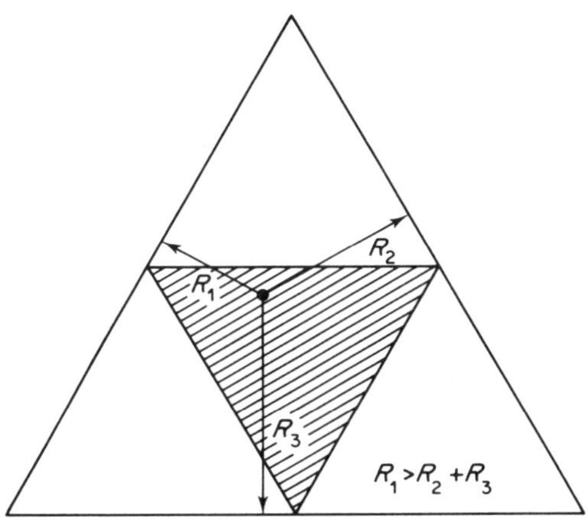

Figure 7.8 Coordinates for a triangular graph. For any point in the triangle $R_1 + R_2 + R_3 = \text{constant} = \text{perimeter of molecular triangle}$. In terms of the S_i variables (4.22), S_1 is constant and S_2 and S_3 are orthogonal coordinates, one of which is parallel to a side of the triangle and the other perpendicular to a side

HNC is not easily studied and was in fact first detected from a radioemission line from the interstellar medium[27] attributed, on the basis of an assumed geometry, to the $J = 1 \leftarrow 0$ rotational transition. It was not until 1975 that it was detected as a gas-phase species in the laboratory. Later Pearson *et al.*[28] reported the

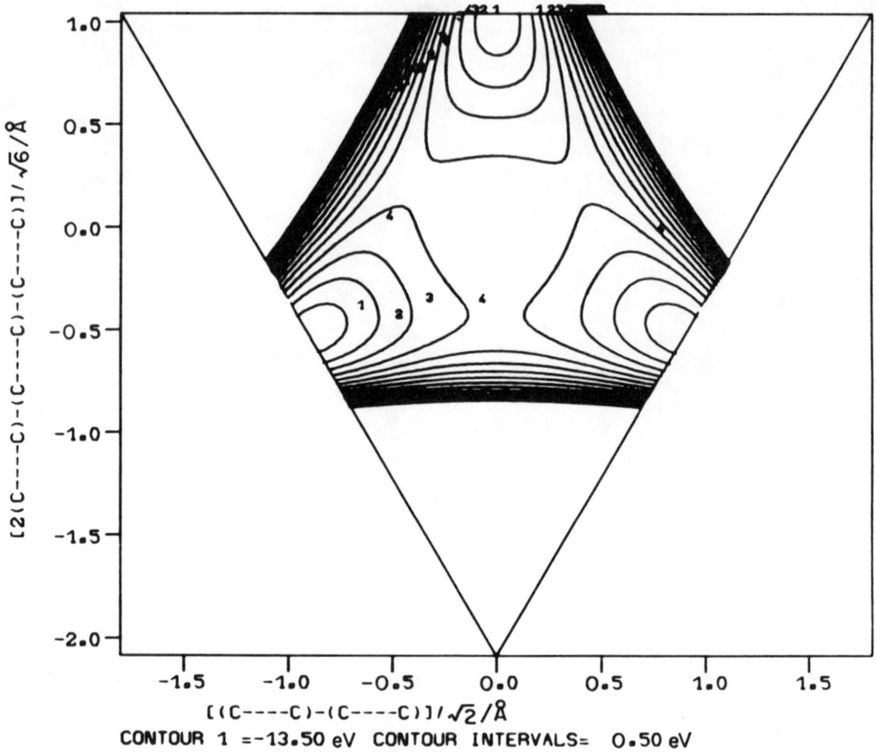

Figure 7.9 A triangular graph for the C_3 potential of Table 7.6 with a fixed perimeter of the triangle of 5.11 Å

millimetre wave spectra of the eight stable isotopic species of HNC, and Maki and Sams[29] determined some of the vibration–rotation constants by high-resolution IR spectroscopy. Creswell and Robiette[30] used these and other data to produce an anharmonic force field and equilibrium geometry of HNC.

The atom plus diatom dissociation limits for the singlet ground state surface are as follows

$$\mathrm{HCN}(\tilde{X}, {}^1\Sigma^+) \rightarrow \begin{array}{l} \mathrm{H}({}^2\mathrm{S}) + \mathrm{CN}(X, {}^2\Sigma^+) \\ \mathrm{C}({}^3\mathrm{P}) + \mathrm{NH}(X, {}^3\Sigma^-) \\ \mathrm{N}({}^4\mathrm{S}) + \mathrm{CH}(a, {}^4\Sigma^-) \end{array} \tag{7.21}$$

Only the CH species is produced in an excited state. The alternative channel, which is the ground-state diatom, CH($^2\Pi$), and excited-state atom, N(^{2}D), is higher in energy for all values of the CH bond length. The N(^{2}D) state is 2.38 eV above N(^{4}S) and calculations[31] show that CH($^4\Sigma^-$) is only 0.7 eV above the ground state at its equilibrium distance and dissociates to the same atomic states, C(^{3}P) + H(^{2}S). The two-body CH potential function can be deduced by fitting to the *ab initio* potential energy curve[32].

The energies, geometries and harmonic force fields of HCN and HNC provide 14 pieces of data in addition to the condition of linearity for both structures[32]. The energy of HNC was taken to be 0.486 eV above that of HCN, which is very close to the value deduced by Maki and Sams[29] from infrared studies of an equilibrium mixture of the two species. The geometry and energy of the transition state have been calculated by Pearson *et al.*[33]: $R_{CH} = 1.171$ Å, $R_{CN} = 1.181$ Å, $R_{NH} = 1.430$ Å, $\Delta E = 2.14$ eV (relative to HCN). These data were used to obtain the initial surface, which was based on an 18-term polynomial in the three-body term in geometry-optimized coordinates S_i chosen, according to the procedure described in Section 4.3, so as to treat the HCN and HNC minima in an even-handed manner[34].

The polynomial in the final frequency-optimized potential was increased in size to include all terms up to quartic, plus one quintic which was found necessary to reproduce the difference in CN bond lengths in the two isomers. The geometry and energy of the transition state were allowed to relax in this iteration and in the final potential had the following values: $R_{CH} = 1.120$ Å, $R_{CN} = 1.139$ Å, $R_{NH} = 1.495$ Å, $\Delta E = 1.51$ eV. The value of R_{CN} is almost certainly slightly too short, but

Table 7.7 Many-body potential for the ground state of HCN (ref. 34). The three-body polynomial is written in S_i variables and the range function (3.27) in ρ_i variables.

Two-body terms:

Species	D_e/eV	R_e/Å	$a_1/\text{Å}^{-1}$	$a_2/\text{Å}^{-2}$	$a_3/\text{Å}^{-3}$
CH	2.8521	1.0823	5.5297	8.7166	5.3082
CN	7.9282	1.1718	5.2448	7.3416	4.9785
NH	3.9938	1.0370	3.0704	0.0	0.0

Three-body term:

$C_1/\text{Å}^{-1}$	1.9076	$C_{1111}/\text{Å}^{-4}$	−0.0367	V^0/eV	−3.0578
C_2	−0.5008	C_{2222}	0.4821	$\gamma_1/\text{Å}^{-1}$	3.9742
C_3	−0.0149	C_{3333}	0.2564	γ_2	4.3688
$C_{11}/\text{Å}^{-2}$	0.6695	C_{1112}	−0.0017	γ_3	1.5176
C_{22}	−1.3535	C_{1122}	−0.2278	R_1^0(CH)/Å	1.9607
C_{33}	−1.0501	C_{1222}	−0.1287	R_2^0(CN)	2.2794
C_{12}	0.2698	C_{1113}	0.1759	R_3^0(NH)	1.8687
C_{13}	−1.1120	C_{1133}	−0.0399		
C_{23}	1.9310	C_{1333}	−0.1447		
$C_{111}/\text{Å}^{-3}$	−0.0877	C_{2223}	−0.3147		
C_{222}	0.0044	C_{2233}	0.1233		
C_{333}	0.0700	C_{2333}	0.3161		
C_{112}	0.0898	C_{1123}	0.0919		
C_{122}	−1.0186	C_{1223}	−0.0954		
C_{113}	−0.0911	C_{1233}	0.1778		
C_{133}	0.0017	$C_{22222}/\text{Å}^{-5}$	−0.1892		
C_{223}	0.4567	$S_1 = 0.4436\rho_1 + 0.6091\rho_2 + 0.6575\rho_3$			
C_{233}	−0.8840	$S_2 = -0.8941\rho_1 + 0.2498\rho_2 + 0.3718\rho_3$			
C_{123}	0.3333	$S_3 = 0.0622\rho_1 - 0.7527\rho_2 + 0.6554\rho_3$			

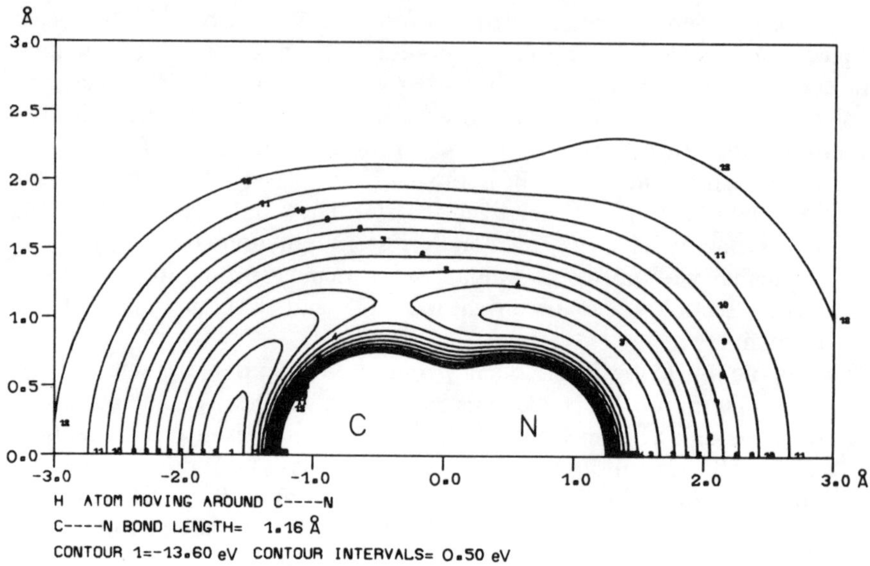

Figure 7.10 Contours of the HCN potential of Table 7.7 for H moving around CN with the C—N bond length fixed at its average value in the equilibrium HCN and HNC structures

the other parameters are probably as reliable as the *ab initio* estimates. The full potential is given in Table 7.7[34].

Figure 7.10 shows energy contours of the potential for a hydrogen atom moving around a CN group whose bond length is the mean of the equilibrium values found in HCN and HNC. The position of the saddle point in this figure is fairly close to the transition state predicted from the potential. The potential also leads to the prediction that there is a shallow minimum on the surface, of depth 0.33 eV, corresponding to a linear hydrogen-bond structure C···HN.

The vibration frequencies of HCN and HNC calculated from this potential, including isotopic substitution of all three atoms, show deviations from experiment of less than 5 cm^{-1} for all levels below 9000 cm^{-1} (ref. 34). The force fields up to quartic terms have also been calculated[33] and compared with the best perturbation force fields of Strey and Mills for HCN (ref. 19) and of Creswell and Robiette[30] for HNC. Whilst there is tolerable agreement in both cases, the differences are larger than was found for SO_2 (Table 7.5), and this emphasizes that for hydrogen-containing species the frequency shifts arising from anharmonicity are so large that they are difficult to calculate accurately by the traditional perturbation method.

References

1. J. Pliva, *Collect. Czech. Chem. Commun.*, **23**, 777, 1839, 1846 (1958).
2. I. M. Mills, Chemical Society Specialist Periodical Report, *Theoretical Chemistry*, Vol. 1, Chemical Society, London, 1974, p. 110.

3. K. S. Sorbie and J. N. Murrell, *Mol. Phys.*, **29**, 1387 (1975).
4. J. F. Ogilvie, *Can. J. Spectrosc.*, **19**, 171 (1973).
5. S. Carter and I. M. Mills, personal communication reported in ref. 6.
6. S. Farantos, E. C. Leisegang, J. N. Murrell, K. S. Sorbie, J. J. C. Texeira-Dias and A. J. C. Varandas, *Mol. Phys.*, **34**, 947 (1977).
7. S. N. Foner and R. L. Hudson, *J. Chem. Phys.*, **36**, 2681 (1962).
8. M. Bottomley, J. N. Bradley and J. R. Gilbert, *Int. J. Chem. Kinet.*, **13**, 957 (1981).
9. A. Gauss, *J. Chem. Phys.*, **68**, 1689 (1978).
10. G. D. Carney, L. L. Sprandel and C. W. Kern, *Adv. Chem. Phys.*, **37**, 305 (1978).
11. S. Carter and N. C. Handy, *Mol. Phys.*, **47**, 1445 (1982).
12. S. Carter, N. C. Handy and B. T. Sutcliffe, *Mol. Phys.*, **49**, 745 (1983).
13. R. J. Whitehead and N. C. Handy, *J. Mol. Spectrosc.*, **55**, 356 (1975).
14. D. E. Milligan and M. E. Jacox, *J. Chem. Phys.*, **38**, 2627 (1963).
15. S. Carter, I. M. Mills, J. N. Murrell and A. J. C. Varandas, *Mol. Phys.*, **45**, 1053 (1982).
16. T. H. Dunning and R. C. Raffenetti, *J. Phys. Chem.*, **85**, 1350 (1981).
17. W. Craven, D. B. Knowles, J. N. Murrell, M. A. Vincent and Z. H. Zhu, unpublished work.
18. R. D. Shelton, A. H. Nielson and W. H. Fletcher, *J. Chem. Phys.*, **21**, 2178 (1953); **22**, 1791 (1954).
19. G. Strey and I. M. Mills, published in ref. 2, p. 155, also in ref. 15.
20. J. N. Murrell, W. Craven and S. C. Farantos, *Mol. Phys.*, **49**, 1077 (1983).
21. E. A. Ballik and D. A. Ramsay, *J. Chem. Phys.*, **31**, 1128 (1959).
22. L. Gausset, G. Herzberg, A. Lagerqvist and A. Rosen, *Discuss. Faraday Soc.*, **35**, 113 (1965); *Astrophys. J.*, **142**, 45 (1965).
23. W. Weltner and D. McLeod, *J. Chem. Phys.*, **45**, 3096 (1966).
24. D. H. Liskow, C. F. Bender and H. F. Schaefer, *J. Chem. Phys.*, **56**, 5075 (1972).
25. J. Romelt, S. D. Peyerimhoff and R. J. Buenker, *Chem. Phys. Lett.*, **58**, 1 (1978).
26. S. Carter, I. M. Mills and J. N. Murrell, *J. Mol. Spectrosc.*, **81**, 110 (1980).
27. D. Buhl and L. E. Snyder, *Bull. Am. Astron. Soc.*, **3**, 388 (1971).
28. E. F. Pearson, R. A. Creswell, M. Winnewisser and G. Winnewisser, *Z. Naturforsch., Teil A*, **31**, 1394 (1976).
29. A. G. Maki and R. L. Sams, *J. Chem. Phys.*, **75**, 4178 (1981).
30. R. A. Creswell and A. G. Robiette, *Mol. Phys.*, **36**, 869 (1978).
31. G. L. Lie, J. Hinze and B. Liu, *J. Chem. Phys.*, **59**, 1872 (1973).
32. J. N. Murrell, S. Carter and A. J. C. Varandas, *Mol. Phys.*, **35**, 1325 (1978).
33. P. K. Pearson, H. F. Schaefer and U. Wahlgren, *J. Chem. Phys.*, **62**, 350 (1975).
34. J. N. Murrell, S. Carter and L. O. Halonen, *J. Mol. Spectrosc.*, **93**, 307 (1982).

Chapter 8

H_3—*a much-studied surface*

The H_3 system has long been a fertile testing ground for the study of elastic, inelastic and reactive scattering processes[1]. It has also been one of the most extensively studied potential energy surfaces by both *ab initio*[2-16] and semi-empirical[17-28] methods. Although most of these studies[2-9,11,14,17-28] refer to the ground-state H_3 molecule, there are also reported calculations on some of its excited states[6,10,12,13,15], many of these stimulated by the recent experimental observation[29-33] of H_3 in an excited Rydberg state. A complete review of the early literature on H_3 has been given by Truhlar and Wyatt[34], and hence we focus here on the more recent work. We further limit the survey of this section to the ground doublet state surface of H_3 which controls the dynamics of hydrogen atoms and molecules in their ground electronic states.

The most extensive and accurate calculations of any polyatomic potential energy surface to date, which are claimed to be within the bounds commonly accepted for chemical accuracy of 1 kcal mol^{-1}, are those of Liu[7] for linear H_3 geometries and of Siegbahn and Liu (SL)[11] for the non-linear geometries of H_3. Using the configuration interaction (CI) method and an extended basis set (4s3p1d) of Slater-type orbitals centred on each hydrogen atom, Liu calculated the saddle point for the exchange process to be a linear symmetric structure of bond length $R_{sp}=1.757a_0$, with a classical barrier height of $\Delta V_{sp}=9.80$ kcal mol^{-1} and an estimated best value of 9.68 ± 0.12 kcal mol^{-1}. For the non-linear calculations, Siegbahn and Liu used a (9s3p1d/4s3p1d) contracted Gaussian basis set. A full CI with this basis for the linear symmetric part of the H_3 surface gave a classical barrier height of 9.84 kcal mol^{-1} and a geometry identical with that obtained from the extended Slater basis set. For the actual non-linear calculations a truncated basis set was selected from a transformation to approximate natural orbitals which, when tested for the linear symmetric calculations, gave a barrier height of 9.90 kcal mol^{-1}.

Besides the classical barrier height and barrier location, the harmonic force constants (F_{11} for symmetric stretching, F_{22} for bending and F_{33} for antisymmetric stretching) of the potential energy surface at the col are also crucially important to ensure a correct description of the reaction kinetics, and further to facilitate a comparison with transition state theory[1]. Table 8.1 compares the

Table 8.1 Properties of various H_3 potential energy surfaces at the linear symmetric saddle point. Except for ΔV_{sp}, which is given in kcal mol^{-1}, all other quantities are in atomic units. Definitions of symbols in text.

Surface	Ref.	R_{sp}	ΔV_{sp}	F_{11}	F_{22}	F_{33}
Ab initio:						
EB–SCF	11	1.724	24.4	0.37	0.017	—
SZ–CCI	2	1.883	23.4	—	—	—
EB–VCSCF	7	1.81	16.6	0.29	—	—
DZ–CCI	3	1.779	15.4	0.29	0.023	−0.047
DZ–CCI	4	1.788	14.0	0.30	—	−0.054
DZP–CCI	4	1.764	11.0	0.31	0.024	−0.061
EB–SD(SCF)	11	1.761	12.0	0.32	0.020	—
EB–SD(VCSCF)	7	1.757	9.9	0.32	—	−0.058
EB–NCCI	7, 11	1.757	9.80*	0.320	0.021	−0.058
Semi-empirical:						
EHP(0)	17	1.775	13.0	0.328	0.020	−0.145
CH(0)	19	1.824	10.8	0.352	0.017	−0.153
PK(2)	18	1.701	9.1	0.364	0.024	−0.124
W(1)	24	1.735	9.8	N.a.	N.a.	N.a.
Least-squares fits:						
VM(19)	21	1.775	10.9	0.322	0.037	−0.062
YL(4)	20	1.742	9.81	0.324	0.030	−0.071
TH(23)	22	1.757	9.80	0.3231	0.0210	−0.0577
W(23)	26	1.756	9.80	N.a.	N.a	N.a
Fit from force field data:						
V(14)	23	1.757	9.68	0.3200	0.0210	−0.0580

* 9.68 ± 0.12.

properties of the various *ab initio* and semi-empirical H_3 potential energy surfaces that we discuss below. Table 8.1 is far from being complete with regard to the work which was carried out prior to 1976 and the reader is referred to the review by Truhlar and Wyatt[34] for further comparisons. The *ab initio* surfaces are ordered from top to bottom roughly in order of increasing complexity, and labelled with SZ, DZ, DZP and EB to denote single-zeta, double-zeta, double-zeta plus polarization and extended one-particle basis sets, respectively. In addition, SCF, VCSCF, CCI, SD(SCF), SD(VCSCF) and NCCI denote self-consistent field, valence configuration self-consistent field, complete CI, all single and double replacement CI with respect to a single root configuration, all single and double replacement CI with respect to a VCSCF root configuration and nearly complete CI, respectively. The semi-empirical and fitted potential energy surfaces are labelled from their authors with the references to the original works being quoted in the second column. For completeness, the number of adjustable parameters is also given, in parentheses, after the authors' initials.

Of the semi-empirical surfaces, those of Ellison *et al.* (EHP)[17] and of Wu

[W(1)][24] are based on the diatomics-in-molecules (DIM) and generalized-DIM formalisms (see Section 3.3), respectively. The former has been calculated *a priori* (i.e. exclusively from available experimental or theoretical data on the diatomic fragments) whereas the latter has one adjustable parameter which has been introduced through an arbitrary geometry function (the so-called *G*-function) which provides a correction to the DIM Hamiltonian. For this Wu adopted the form $G=1-cS_1S_2S_3$, where S_i $(i=1\text{–}3)$ are the three overlap integrals, and determined the adjustable parameter c by fitting Liu's *ab initio* result for the classical barrier height. It should be noted that DIM theory (with neglect of orbital overlap) leads to the London equation (3.11) in the case of a triatomic system formed from 1s hydrogen atoms[35].

Another semi-empirical DIM surface CH (ref. 19), which also relies on an *a priori* evaluation of the DIM equation, has saddle point properties which differ slightly from those of the EHP surface; this is due to the use of slightly different diatomic potentials by the two groups. A much-used potential by Porter and Karplus (PK)[18] is based on the semi-empirical basis set valence bond method in which overlap and three-centre integrals are retained. The PK surface has two adjustable parameters which were determined by fitting the classical barrier height to the experimental activation energy for the hydrogen atom exchange reaction while making the characteristic bond length of the transition state agree as far as possible with the *ab initio* value then available. At D_{3h} geometries the PK function reduces to a simple expression which contains overlap but no adjustable parameters. This result has recently been used by Eaker and Allard[28] to attempt a theoretical derivation of the functional form of the geometry function of GDIM theory by comparing the DIM and PK expressions for these geometries.

With the appearance of accurate *ab initio* calculations on H_3, the emphasis moved to model potentials capable of matching accurately the *ab initio* data. The topographical similarities between the semi-empirical and *ab initio* surfaces near the saddle point region suggested the semi-empirical functions as good candidates for that task. One such function is that of Yates and Lester (YL)[20], which is based on a modified form of the PK surface. The YL function contains four adjustable parameters which were determined by a least-squares fit to Liu's CI energies for linear geometries. The YL potential at non-linear geometries is therefore determined exclusively from the adopted analytical representation. It is now known that the YL potential is much too repulsive at non-linear geometries. In addition, Choi and Tang[36] noticed that the YL function reduces to the attractive part of the diatomic curve for the ground singlet state of H_2 when the other atom moves to infinity, the repulsive part of the diatomic potential being replaced by zero. Further, they did not find a simple way of altering this peculiar behaviour of the asymptotic form of the YL function.

The best available three-dimensional *ab initio* energies prior to those of Siegbahn and Liu were those of Shavitt *et al.* (SSMK)[4] and of Porter *et al.* (PSK)[5]. These were fitted by Varandas and Murrell (VM)[21] to a sum of two- and three-body terms using the type of functions described in Section 3.5. The two-body potentials for the ground singlet state of H_2 were represented by Extended

Rydberg functions (1.19) fitted to the *ab initio* data[4] on the diatomic fragments. The three-body term was expressed in the integrity basis for three variables(4.27), and to allow for the fact that the H_3 ground state is doubly degenerate for D_{3h} geometries, and hence has a conical intersection characteristic of a Jahn–Teller system at those geometries (Section 2.2) non-analytic powers of $q_r = (S_2^2 + S_3^2)^{1/2}$ (ref. 5) were also used. The resulting three-body term took the form

$$V^{(3)}(R_1, R_2, R_3) = (\alpha_0 + \alpha_1 S_1 + \beta_1 q_r + \alpha_2 S_1^2 + \beta_2 S_1 q_r + \alpha_3 q_r^2 + \cdots$$
$$+ \alpha_{10} q_r^4 + \beta_6 S_1 q_r^3 + \beta_7 q_r q_s^3)[1 - \tanh(\gamma S_1/2)] \qquad 8.1$$

where α_i and β_i are the coefficients of the analytic and non-analytic terms, respectively, and $q_s^3 = S_3^3 - 3S_2^2 S_3$ represents the cubic integrity basis polynomial (4.25). Five of the α_i (those involving the terms S_1^n, including α_0) were determined by fitting exactly five points of the PSK *ab initio* curve for D_{3h} geometries, in addition to the requirement that the D_{3h} minimum should be at $R = 1.965a_0$ (ref. 5). The non-analytic coefficient β_1 was constrained to have the value quoted by Porter *et al.*[5] for the slope of the Jahn–Teller conical intersection, while the remaining α_i and β_i coefficients were chosen from a least-squares fitting procedure to the SSMK *ab initio* energies at bond angles of 120°, 150° and 180°. The r.m.s.d. of the VM fit is 1.26 kcal mol^{-1}, with the maximum error relative to three hydrogen atoms of 2%, this being at a high energy part of the surface (typical errors were 0.3% over most of the fitted points.

It was not until Siegbahn and Liu computed the non-linear part of the H_3 surface that a better analytical least-squares function could be obtained. One such fit, which was published simultaneously with the Siegbahn and Liu paper, is that of Truhlar and Horowitz (TH)[22]. The TH surface is based on the London form (3.11), with the triplet state potential for the diatomic fragments being taken as an arbitrary fitting function; this is acceptable since it does not represent the asymptotic limit in adiabatic scattering processes between an H atom and an H_2 molecule in their ground states. Further flexibility was introduced by the addition of three-body correction terms. The TH function has eight linear and fifteen non-linear parameters, and fits the 256 reported SL points with an r.m.s.d. of 0.17 kcal mol^{-1} and a maximum deviation of 0.55 kcal mol^{-1}. An improvement to this least-squares fit has recently been reported by Johnson[27], who noticed that there were discontinuous derivatives in the TH function for certain isosceles triangular geometries above 236 kcal mol^{-1} (relative to the zero of energy of the reactants as zero).

The TH surface has been used extensively for studies of the $H + H_2$ dynamics. However, notwithstanding its high accuracy, recent dynamical studies by Pollack and Wyatt[37] suggest that it may not be sufficiently accurate in regions of the surface where the density of fitted *ab initio* points is low. A more recent least-squares fit to the SL *ab initio* energies has been performed by Wu[26] using a GDIM functional form. Using a total of 23 parameters in the geometry function, Wu reported an r.m.s.d. of 0.33 kcal mol^{-1} and a maximum deviation of 0.85 kcal mol^{-1} from the available 256 *ab initio* points.

The results in Table 8.1 indicate that the classical barrier height converges more slowly at the *ab initio* level than the other saddle point properties. This suggests that it will not be generally possible to obtain an accurate surface for dynamical studies from a least-squares fitting procedure to *ab initio* points without first scaling them to give a better estimate for the classical barrier height. An alternative approach is to parameterize the potential energy function from the best available estimates for the geometry, energy and force constants of the saddle point. Varandas (V)[23] has shown how this may be done using a LEPS-type form defined by equation (3.11) in which the Q_{ij} and A_{ij} integrals are evaluated from equation (3.4) with $S_{ij}=0$. It is this approach that we examine next.

In Chapter 7 we discussed the problem of fitting potentials to spectroscopic data and, in Section 7.1 in particular, of fitting the potential to the energy, geometry and force constants of a molecular species. There are no differences of principle between this and fitting the potential to the energy, geometry and force constants of a saddle point, as this is also a stationary point on the surface. Moreover, fitting the properties of a saddle point for a linear structure, as for H_3, is an equivalent problem to fitting the properties of a stable linear species, which was discussed in Section 7.4. If the potential of a triatomic species is to be expressed in terms of the three internuclear distances R_i, then one needs first to establish relationships between the derivatives with respect to R_i and the derivatives with respect to the internal valence coordinates, two bond lengths and the included angle. These relationships are given in Appendix 1a.

Varandas[23] has shown that the energy, geometry and quadratic force constants of the H_3 saddle point can be used to determine six parameters in the triplet component (3.4) of the LEPS-type function; the detailed relationships are given in his paper. For the triplet state potential he used a hybrid form:

$$V_T(R)=\left(\sum_{i=0}^{5} b_i R^i\right)\exp(-\gamma_1 R), \qquad R\leqslant 2R_{sp} \tag{8.2}$$

$$V_T(R)=\left(\sum_{i=0}^{5} c_i R^i\right)\exp(-\gamma_2 R)+Q_{as}(R), \quad R>2R_{sp} \tag{8.3}$$

where $Q_{as}(R)$ represents the leading terms of the multipolar expansion of the long-range dispersion energy (asymptotic limit of the Coulomb energy) for the interaction of two hydrogen atoms, i.e.

$$Q_{as}(R)=-\sum_{L=3}^{5} C_{2L}R^{-2L} \tag{8.4}$$

However, to determine uniquely the b_i ($i=0$–5) coefficients one needs to specify the energy of the diatomic triplet state curve at $R=2R_{sp}$. The actual value that was considered is about 0.5 kcal mol^{-1} higher than the accurate Kolos and Wolniewicz (KW)[38] *ab initio* result for the lowest triplet state of H_2, and was chosen to give the spherically averaged component of the H–H_2 interaction in good agreement with the molecular beam estimate of Gengenbach *et al.*[39]. The range determining coefficient γ_1 was left as a freely adjustable parameter, and chosen to make the triplet state function (8.2) mimic as closely as possible the KW *ab initio* triplet state curve at short distsnces. The six coefficients c_i in the outer part of the effective triplet state curve were obtained from the condition that the complete potential should be continuous and have continuous first and second derivatives at $R=2R_{sp}$, in addition to the requirement that it reproduced the van der Waals attributes (distance of energy zero σ, location of the van der Waals minimum R_m and well depth ε) of the KW curve. As for γ_1, the range parameter of the outer potential (γ_2) was left as a freely adjustable parameter and set equal to the value of the range parameter for the ground singlet state diatomic potential. To represent this the following Hybrid Extended Rydberg[40] form was adopted:

$$V_S(R) = -D_e\left(1+\sum_{i=1}^{5} a_i r^i\right)\exp(-a_1 r), \quad R \leqslant R_0 \qquad 8.5$$

$$V_S(R) = Q_{as}(R) + A_{as}(R), \qquad R > R_0 \qquad 8.6$$

where $r=R-R_e$, $Q_{as}(R)$ is defined by equation (8.4), and

$$A_{as}(R) = -kR^a \exp(-a'R) \qquad 8.7$$

is the leading term in the asymptotic expansion of the exchange interaction between two H atoms; values of k, a and a' for the hydrogen and alkali metal atoms are given in reference 41. Following the general approach described in Section 1.2, the a_i ($i=1$–5) coefficients in the inner potential were determined from the force constants up to the quartic term, and also from the requirement that the complete potential should be continuous and have continuous first derivative at R_0. A useful criterion to define the latter is that of Le Roy[42] (5.8) for the point at which the asymptotic expansion (8.4) breaks down due to strong orbital overlap. The complete H_2 potential, together with the potential energy surface for ground-state H_3, are numerically defined in Table 8.2. The former provides an accurate representation of the ground singlet state diatomic curve over all the attractive region of the potential, which is important for the purpose of studying atom–diatom dynamics, particularly when the diatomic molecule is vibrationally excited. This requirement was also ensured by Truhlar and Horowitz[22], who employed a cubic spline representation of the Kolos and Wolniewicz[38] *ab initio* curve for the ground singlet state of H_2, and by Wu[25,26], who adopted the Hybrid Extended Rydberg form [equations (8.5) and (8.6)].

Although the LEPS-type function derived in this way shows very good agreement with the *ab initio* energies of Liu for the linear geometries of H_3, it deviates from the *ab initio* surface of Siegbahn and Liu with decreasing values of the included angle, up to a maximum deviation for the equilateral triangular geometries. It was found that a simple way to correct this was by adding the term

$$V_{ad}(R_1,R_2,R_3) = \left\{1+\sum_{i=1}^{3}\left[(R_i^2-R_j^2-R_k^2)/2R_jR_k\right]\right\}^4 (d_0+d_1P)\exp(-1.09P) \qquad 8.8$$

Table 8.2 An accurate potential energy surface for H_3 from force field data[23]. All quantities are in atomic units.

$$V(R_1, R_2, R_3) = V_{\mathrm{LEPS}}(R_1, R_2, R_3) + V_{\mathrm{ad}}(R_1, R_2, R_3)$$

$$V_{\mathrm{LEPS}}(R_1, R_2, R_3) = \sum_{i=1}^{3} Q_i - \left[\frac{1}{2}\sum_{i>j=1}^{3} (A_i - A_j)^2\right]^{1/2}$$

$$Q_i(R) = [V_{\mathrm{S}}(R) + V_{\mathrm{T}}(R)]/2$$

$$A_i(R) = [V_{\mathrm{S}}(R) - V_{\mathrm{T}}(R)]/2$$

$$V_{\mathrm{S}}(R) = -0.17457(1 + 1.73034r + 0.43835r^2 + 0.24517r^3 - 0.12884r^4 + 0.013723r^5)$$
$$\times \exp(-1.73034r) \text{ with } r = R - 1.40059, \quad \text{for } R \leqslant 7.0a_0$$

$$V_{\mathrm{S}}(R) = Q_{\mathrm{as}}(R) - 0.8205R^{5/2}\exp(-2R), \quad \text{for } R > 7.0a_0$$

$$Q_{\mathrm{as}}(R) = -6.49903R^{-6} - 124.3991R^{-8} - 3285.828R^{-10}$$

$$V_{\mathrm{T}}(R) = 1.91739(1 - 1.46746R + 0.99100R^2 - 0.35342R^3 + 0.063750R^4 - 0.0045724R^5)$$
$$\times \exp(-0.2R), \quad \text{for } R \leqslant 3.514a_0$$

$$V_{\mathrm{T}}(R) = 329.08328(1 - 0.85572R + 0.30000R^2 - 0.051607R^3 + 0.0043686R^4 - 0.00014582R^5)$$
$$\times \exp(-1.73034R) + Q_{\mathrm{as}}(R), \quad \text{for } R > 3.514a_0$$

$$V_{\mathrm{ad}}(R_1, R_2, R_3) = \left[1 + \sum_{i=1}^{3} (R_i^2 - R_j^2 - R_k^2)/(2R_jR_k)\right]^4 [22.69093 - 26.83568(R_1 + R_2 + R_3)]$$
$$\times \exp[-1.09(R_1 + R_2 + R_3)]$$

where $P = (R_1 + R_2 + R_3)$ and d_0 and d_1 are parameters which are determined from the condition that the minimum of the D_{3h} curve should be at $R = 1.981a_0$ with an estimated best energy of 62.85 kcal mol^{-1} above the atom–diatom dissociation limit[23]. While this correction term brings the complete surface into very close agreement with the SL *ab initio* data at equilateral and near equilateral triangular geometries, it has practically no effect near the linear geometries. In particular, the force field of the LEPS saddle point is left unaltered.

Figure 8.1 shows contour plots of the V potential defined in Table 8.2 for $\alpha = 180°$, and Figure 8.2 shows the energy variation along the minimum energy path in the (R_1, R_2) plane and compares it with the same attribute of other surfaces. For $\alpha = 180°$, the V potential path deviates by no more than 0.2 kcal mol^{-1} from the *ab initio* curve of Liu, and similar agreement is expected with the TH least-squares fit for other angles as shown by the point marked TH in Figure 8.2. The similarity of the V and TH surfaces at short distances is further corroborated from quasi-classical reaction dynamics studies of $H + H_2$ with H_2 in the ground ($v = 0$) and first ($v = 1$) excited vibrational states[43]. In particular, the calculated $v = 1$ cross-sections and rate constants from the V surface are found to be in good agreement with those reported[44] for the TH surface, thus supporting previous theoretical results [for a recent account of the $H + H_2$ ($v = 1$) dynamics, see ref. 45] while challenging the available experimental data on the cooling of vibrationally

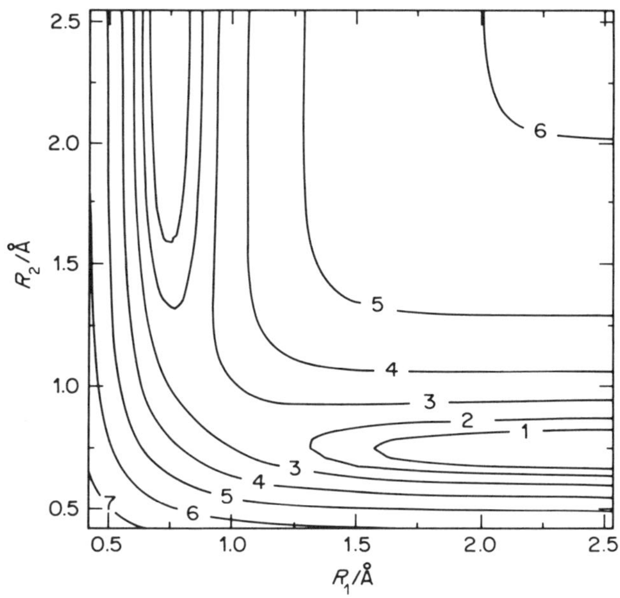

Figure 8.1 Bond stretching contour plot of the H_3 surface defined in Table 8.2 for α fixed at 180°. Contours which are numbered with an integer (n) correspond to curves of constant potential energy (in eV, measured with respect to the energy of the reactants as zero) defined by $0.130(2^{n-1})$

excited H_2 ($v=1$) in collisions with H atoms[46-48] which strongly favour the vibrationally adiabatic channel:

$$H + H_2\ (v=1) \rightarrow H_2\ (v=1) + H$$

over its non-adiabatic counterpart

$$H + H_2\ (v=1) \rightarrow H_2\ (v=0) + H$$

Figure 8.3 shows contours of the H_3 potential for $\alpha=60°$. We note that the H_3 ground-state surface is the lower sheet of a conical intersection and hence displays a ridge along the diagonal ($R_1=R_2$) of this figure where the lower [$V_-(R_1, R_2, R_3)$] and upper [$V_+(R_1, R_2, R_3)$] sheets intersect each other. We further note that this Jahn–Teller instability is inherent in the three-body part of the potential, as shown in Figure 8.4, which displays contours of the three-body term for a fixed perimeter ($7.027a_0$) of the triangle defined by the positions of the H atoms which corresponds to the value at the linear saddle point. Hence, the sharply peaked point in this plot corresponds to the D_{3h} geometry defined by the symmetry coordinates ($S_1=0.626a_0$, $S_2=S_3=0$), while the points indicated by the crosses correspond to the linear saddle points.

Figure 8.5 shows contours of the H_3 potential energy surface for an H atom moving around an equilibrium H_2 molecule whose centre of mass is fixed at the origin. We note that the small van der Waals well is deeper at C_{2v} geometries, in

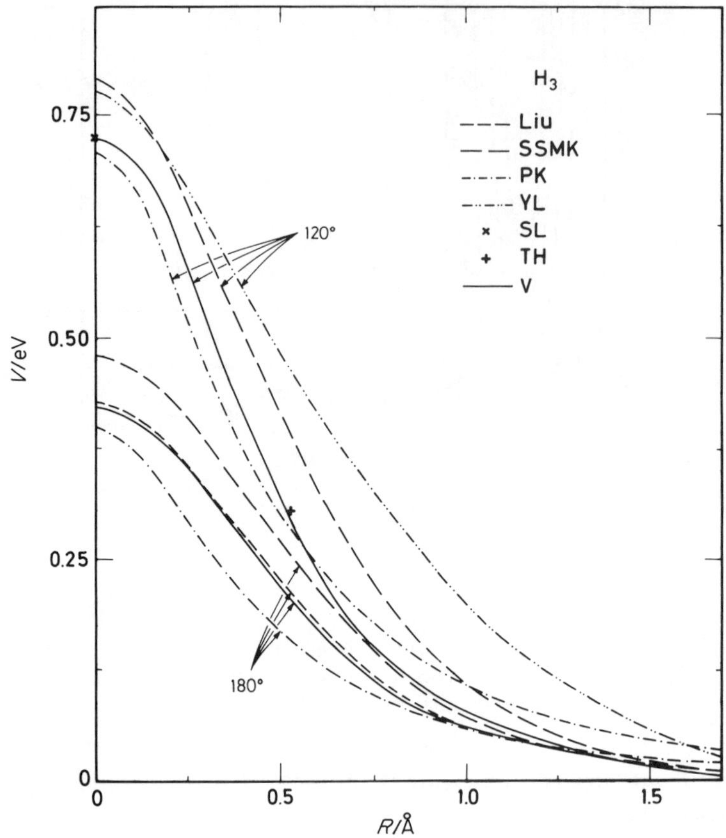

Figure 8.2 Energy profiles along the minimum energy path in the (R_1, R_2) plane for fixed angles $\alpha = 120°$ and $180°$

agreement with the best empirical estimates[39], but in disagreement with the prediction from the TH surface which places it at $R_{HH} = 1.40a_0, r = 6.47a_0, \theta = 47°$. In contrast with the TH and V surfaces, most other semi-empirical surfaces are entirely repulsive at large r.

Two important quantities in the study of H–H_2 scattering processes are the spherically averaged, $V_0(R_{HH}, r)$, and leading anisotropic, $V_2(R_{HH}, r)$, components of the H_3 surface with the H_2 internuclear separation being frozen at its equilibrium geometry. In non-reactive H–H_2 collisions, $V_0(R_{HH}, r)$ controls how closely on average the atom and molecule can approach each other, while $V_2(R_{HH}, r)$ gives rise to torques on H_2 which can induce rotational excitations. The $V_0(R_{HH}, r)$ and $V_2(R_{HH}, r)$ potentials can be obtained, exactly, from the complete surface by integration or, approximately, from a least-squares fitting procedure to *ab initio* points using a truncated Legendre series expansion (4.2). A graphical comparison of the $V_0(R_{HH}, r)$ and $V_2(R_{HH}, r)$ curves obtained from the most popular *ab initio* and semi-empirical surfaces has been given in reference 23.

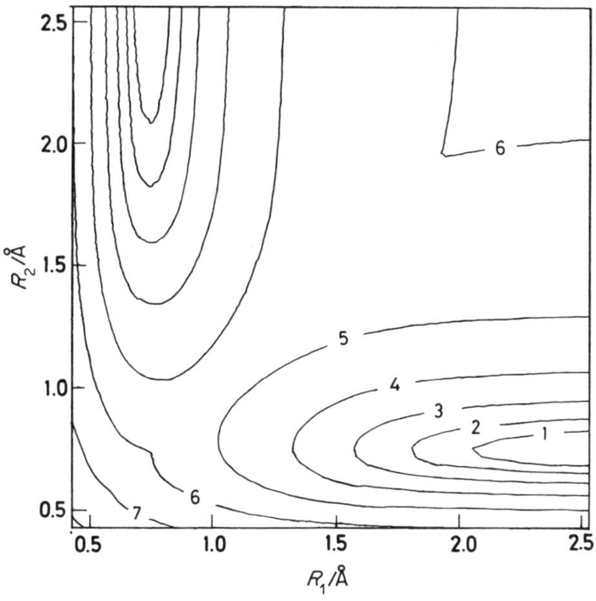

Figure 8.3 Bond stretching contour plot of the H_3 surface defined in Table 8.2 for α fixed at 60°. Contours as in Figure 8.1

While the V and TH surfaces agree well for $V_0(R_{HH}, r)$, they show significant differences for $V_2(R_{HH}, r)$. For example, the leading anisotropic component of the V surface changes sign three times: it is positive for $r \leqslant 2.1a_0$, negative for $2.1 < r < 2.8a_0$, positive for $2.8 < r < 7.4a_0$ and finally becomes negative again for $7.4a_0 < r < < \infty$ with a minimum about 2 cal mol^{-1} deep occurring at $r \approx 8a_0$. The TH surface also shows three changes of sign but they are, in the above order, at $r = 2.0$, 5.1 and $6.2a_0$.

At large separations the H–H_2 interaction potential is dominated by the dispersion energy which can be approximated by

$$V^{\text{disp}}(R_{HH}, r, \theta) = V_0^{\text{disp}}(R_{HH}, r) + V_2^{\text{disp}}(R_{HH}, r)$$

$$= -\sum_{L=3} C_{2L} r^{-2L}[1 + \Gamma_{2L} P_2(\cos\theta)] \qquad 8.9$$

where Γ_{2L} are the anisotropy coefficients which determine the energy difference between the linear and perpendicular geometries. One can deduce from equation (8.9) that

$$V^{\text{disp}}(R_{HH}, r, \pi/2) - V^{\text{disp}}(R_{HH}, r, 0) = (3/2) \sum_{L=3} C_{2L} \Gamma_{2L} r^{-2L} \qquad 8.10$$

which shows that, at very large r, the linear approach of the incoming atom is preferred if the Γ_{2L} are positive. Although reliable estimates of C_6 have been known for some time[34], it is only recently that accurate values of C_8 and C_{10} (in

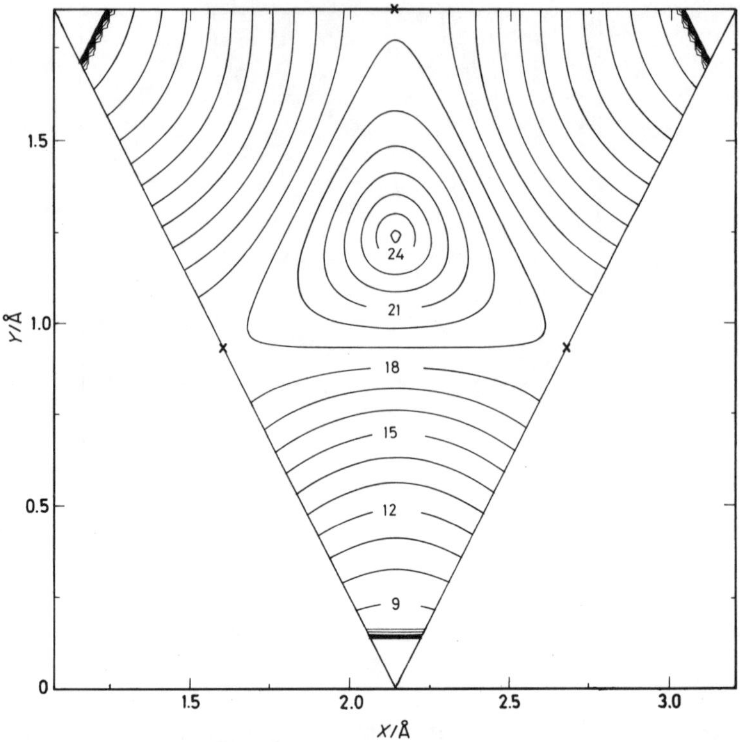

Figure 8.4 The three-body part of the H_3 surface of Table 8.2 for a fixed perimeter of the triangle defined by the three H atoms: $R_1+R_2+R_3=3.719$ Å. Contours, which are labelled with an integer (*n*), are shown for all positive values of the three-body energy. The energies themselves (in eV) are given by 0.272*n*. $X=(2R_2+R_3)/3^{1/2}$, $Y=R_3$. The points × are the geometries corresponding to the linear saddle points

addition to C_6) have been obtained[49]. These are, in atomic units, 8.813, 163.87 and 4023.2 for C_6, C_8 and C_{10}, respectively. For the anisotropy coefficients the only published values are for Γ_6 with typical values being close to 0.11. A calculation of Γ_8 has been made using the procedure recently employed by Tang and Toennies[50] for Ne–H_2, which consists in expressing Γ_8 in terms of the frequency-dependent polarizabilities for H and H_2; the result is $\Gamma_8=0.265$ (ref. 51). We note that a similar calculation for C_6, C_8 and Γ_6 gives $8.816E_ha_0^6$, $162.95E_ha_0^8$ and 0.108, respectively, in good agreement with the best available estimates for these coefficients. If one now makes the assumption[50] that the anisotropy of the higher-order terms is identical with that of C_8, one obtains $\Gamma_{10}=0.265$. Thus the values of Γ_8 and Γ_{10} obtained in this way for H–H_2 are close to those reported for He–H_2 (ref. 52) and Ne–H_2 (ref. 50), which are of the order of 0.3.

Experimentally, information on $V_0(R_{HH}, r)$ and $V_2(R_{HH}, r)$ comes from the inversion of transport and cross-section measurements. However, transport data depend only weakly on the H–H_2 interaction potential, as has been demonstrated

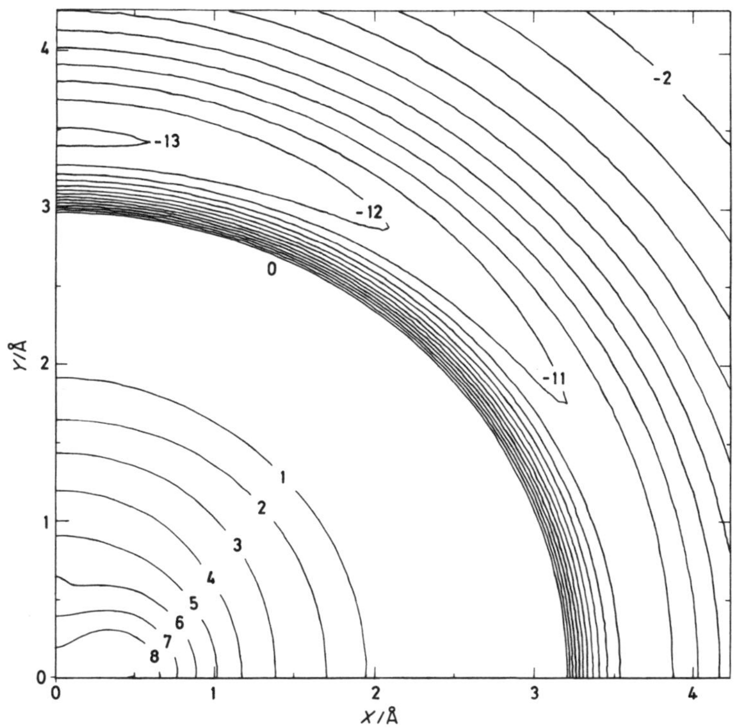

Figure 8.5 Numbered (n) contours of the H_3 potential energy surface defined in Table 8.2 for an H atom moving around an H_2 molecule fixed at its equilibrium geometry and lying along the X axis with the centre of mass at the origin. The contours (in eV) are given by $0.000217n$ in the attractive region of the potential, and in the repulsive region they are given by $0.130(2^{n-1})$

by the wide range of potentials which have been found to provide an equally good representation of the experimental data. Truhlar and Wyatt[34] provided a compilation and critical analysis of such potentials (see also ref. 53). More consistent information is obtained from the molecular beam data[39,54–58], as shown by the characteristic van der Waals parameters which have been reported for the H–H_2 isotropic interaction potential. A comparison of such characteristic parameters for the most recent potentials is given in Table 8.3.

Although the H_3 ground-state surface has been frequently studied by *ab initio* methods, most calculations concentrated on the region close to the linear saddle point for the exchange reaction. Specific *ab initio* calculations for studying the anisotropy of the H–H_2 interaction were carried out by Norbeck and co-workers[8,9] using an optimized (2s, 1p) Slater basis set. However, these calculations did not extend to the van der Waals region, and hence did not allow a comparison with experiment on that region. Varandas and Tennyson[14] performed SCF calculations of $V_0(R_{HH}, r)$ and $V_2(R_{HH}, r)$ for H approaching an H_2 molecule with its bond length fixed at the equilibrium geometry, using the

Table 8.3 Characteristic parameters for the van der Waals well of the H–H_2 isotropic interaction potentials.

Ref.	Model	σ/a_0	r_m/a_0	ε/kcal mol^{-1}
Gengenbach *et al.*[39]	BMSV[a]	5.93	6.71	0.054
Welz[54]	LJ(12, 6)[b,c]	5.93	6.65	0.046
Welz[54]	exp (16.8, 6)[b,d]	5.93	6.56	0.060
Hishinuma[55]	exp (12, 6)[d]	5.63	6.33	0.055
Hishinuma[57]	MBMD[e]	5.65	6.48	0.048
Hishinuma[58]	MBMD[e]	5.65	6.48	0.047
Torello and Dondi[56]	MBMD[e]	5.66	6.56	0.042
Truhlar and Horowitz[22]	[f]	5.78	6.59	0.036
Varandas[23]	[g]	5.76	6.66	0.060
Varandas and Tennyson[14]	MBMD[h]	5.73	6.58	0.049

[a] Three-region Born–Mayer–spline–van der Waals ($-C_6r^{-6}-C_8r^{-8}$) potential.
[b] For $r\leqslant\sigma$ $V(r)$=potential of reference 39.
[c] Lennard–Jones (6–12) potential, $V(r)=\varepsilon[(r_m/r)^{12}-2(r_m/r)^6]$.
[d] Modified Buckingham (6–exp) potential, $V(r)=[\varepsilon/(1-6/\alpha)]\{(6/\alpha)\exp[(1-r/r_m)]-(r_m/r)^6\}$.
[e] Modified Born–Mayer–dispersion potential, $V(r)=A\exp(-br)-(C_6r^{-6}-C_8r^{-8}-C_{10}r^{-10})\exp(-\rho r_m/r-1)^2$.
[f] Isotropic component of TH least-squares surface.
[g] Isotropic component of V surface; see Table 6.3.
[h] Modified Born–Mayer–dispersion potential, equation (6.27).

extended (4s3p1d) Gaussian basis set of Siegbahn and Liu. In addition, they carried out SCF CI calculations for the C_{2v} and $C_{\infty v}$ geometries of H–H_2 using all single and double replacements from the ground-state SCF root configuration, which amounted to 1694 and 1946 configurations, respectively. From this basis and limited CI, only about half of the dispersion energy predicted from the semi-empirical multipole expansion was obtained. However, an accurate analytic estimate of V_0 has been obtained by adding to the short-range near-Hartree–Fock curve (which was fitted to a modified Born–Mayer form) a function which approximately fitted the correlation energy at small values of r, while reducing at large r to the correct long-range multipolar expansion. The recommended form of the complete potential is[14]

$$V_0(R)=A\exp(-e_1r-e_2r^2)-\left(\sum_{L=3}^{5}C_{2L}r^{-2L}\right)[1-\exp(-h_1r^2-h_2r^4)] \qquad 8.11$$

with $A=0.64299E_h$, $e_1=0.97919a_0^{-1}$, $e_2=0.05775a_0^{-2}$, $h_1=-0.003625a_0^{-2}$ and $h_2=0.0012572a_0^{-4}$. The C_{2L} dispersion coefficients were taken from Mayer[49]. We note that a calculation[53], using the first-order approximation of the classical Chapman–Enskog theory, of the binary diffusion coefficient of atomic and molecular hydrogen at the temperature of 294 K based on the isotropic potential (8.11) has given $D_{12}=2.07$ cm^2 s^{-1}, and to second-order 2.09 cm^2 s^{-1}, which is in excellent agreement with the most recent and accurate experimental result of

$D_{12} = 2.09 \pm 0.02$ cm^2 s^{-1} (ref. 59). Such an agreement therefore suggests that any error due to the neglect of orientation-dependent contributions is small at that temperature. More recently, Tennyson[60] has suggested a functional form similar to equation (8.11) for $V_2(R_{HH}, r)$ but with the summation restricted to the leading anisotropy term $\Gamma_6 C_6$. Using the new $V_2(R_{HH}, r)$ in conjunction with equation (8.11), he then performed nuclear dynamical calculations on the H–H_2 van der Waals molecule. He found H_3 to have only one weakly bound state, while similar calculations on other isotopic variants implied several bound states.

A more recent calculation of $V_0(R_{HH}, r)$ and $V_2(R_{HH}, r)$ for the H + H_2 surface at the near Hartree–Fock level has been reported by Lewchenko *et al.*[16] using a contracted (11s4p1d/6s4p1d) Gaussian basis set, and considering H_2 geometries between $R_{HH} = 1.2$ and $1.6a_0$. For $R_{HH} = 1.40a_0$, these calculations give energy values which are lower than those of Varandas and Tennyson[14] by less than $10^{-4}E_h$ for $\infty > r > 2.5a_0$. In addition, a vibrational averaging calculation has been found to produce no effect on V_0, but a substantial effect on V_2. This result has been conjectured[16] to be responsible for vibrational effects on the calculation of rotational cross-sections for collision energies greater than 0.3 eV.

References

1. D. G. Truhlar and R. E. Wyatt, *Annu. Rev. Phys. Chem.*, **27**, 1 (1976).
2. I. Shavitt, *J. Chem. Phys.*, **31**, 1359 (1959).
3. S. F. Boys and I. Shavitt, *Wisc. Naval Res. Lab. Tech. Rep.*, WISC-AF-13 (1959).
4. I. Shavitt, R. M. Stevens, F. L. Minn and M. Karplus, *J. Chem. Phys.*, **48**, 2700 (1968).
5. R. N. Porter, R. M. Stevens and M. Karplus, *J. Chem. Phys.*, **49**, 5163 (1968).
6. E. Frenkel, *Z. Naturforsch., Teil A*, **25**, 1265 (1970).
7. B. Liu, *J. Chem. Phys.*, **58**, 1925 (1973).
8. J. M. Norbeck and P. R. Certain, *J. Chem. Phys.*, **63**, 4127 (1975).
9. J. M. Norbeck, P. R. Certain and K. T. Tang, *J. Chem. Phys.*, **63**, 590 (1975).
10. J. N. Murrell, A. J. C. Varandas and M. F. Guest, *Mol. Phys.*, **31**, 1129 (1976).
11. P. Siegbahn and B. Liu, *J. Chem. Phys.*, **68**, 2457 (1978).
12. H. F. King and K. Morokuma, *J. Chem. Phys.*, **71**, 3213 (1979).
13. M. Jungen, *J. Chem. Phys.*, **71**, 3540 (1979).
14. A. J. C. Varandas and J. Tennyson, *Chem. Phys. Lett.*, **77**, 151 (1981).
15. Ch. Nager and M. Jungen, *Chem. Phys.*, **70**, 189 (1982).
16. V. Lewchenko, G. C. Hancock and P. R. Certain, *J. Chem. Phys.*, **76**, 3119 (1982).
17. F. O. Ellison, N. T. Huff and J. C. Patel, *J. Am. Chem. Soc.*, **85**, 3544 (1963).
18. R. N. Porter and M. Karplus, *J. Chem. Phys.*, **40**, 1105 (1964).
19. J. K. Cashion and D. R. Herschbach, *J. Chem. Phys.*, **40**, 2358 (1964).
20. A. C. Yates and W. A. Lester, *Chem. Phys. Lett.*, **24**, 305 (1974).
21. A. J. C. Varandas and J. N. Murrell, *Faraday Discuss. Chem. Soc.*, **62**, 92 (1977).
22. D. G. Truhlar and C. J. Horowitz, *J. Chem. Phys.*, **68**, 2466 (1978).
23. A. J. C. Varandas, *J. Chem. Phys.*, **70**, 3786 (1979).
24. A. A. Wu, *Mol. Phys.*, **38**, 843 (1979).
25. A. A. Wu, *Mol. Phys.*, **39**, 1287 (1980).
26. A. A. Wu, *Mol. Phys.*, **42**, 379 (1981).
27. B. R. Johnson, *J. Chem. Phys.*, **74**, 754 (1981).
28. C. W. Eaker and L. R. Allard, *J. Chem. Phys.*, **74**, 1821 (1981).
29. M. Volger, *Phys. Rev., A*, **19**, 1 (1979).
30. G. Herberg, *J. Chem. Phys.*, **70**, 4806 (1979).

31. I. Dabrowski and G. Herzberg, *Can. J. Phys.*, **58**, 1238 (1980).
32. G. Herzberg and J. K. G. Watson, *Can. J. Phys.*, **58**, 1250 (1980).
33. G. Herzberg, H. Lew, J. J. Sloan and J. K. G. Herzberg, *Can. J. Phys.*, **59**, 428 (1981).
34. D. G. Truhlar and R. E. Wyatt, *Adv. Chem. Phys.*, **36**, 141 (1977).
35. B. T. Pickup, *Proc. R. Soc. London, Ser. A*, **333**, 69 (1973).
36. B. H. Choi and K. T. Tang, *J. Chem. Phys.*, **63**, 2854 (1975).
37. E. Pollack and R. E. Wyatt, *J. Chem. Phys.*, **78**, 4464 (1983).
38. W. Kolos and L. Wolniewicz, *J. Chem. Phys.*, **43**, 2429 (1965).
39. R. Gengenbach, Ch. Hahn and J. P. Toennies, *J. Chem. Phys.*, **63**, 4127 (1975).
40. A. J. C. Varandas, *J. Chem. Soc., Faraday Trans. 2*, **76**, 129 (1980).
41. C. Herring and M. Flicker, *Phys. Rev. A*, **134**, 362 (1964).
42. R. J. Le Roy, Chemical Society Specialist Periodical Report, *Molecular Spectroscopy*, Vol. 1, Chemical Society, London, 1973, p. 113.
43. A. J. C. Varandas, *Chem. Phys.*, **69**, 295 (1982).
44. H. R. Mayne and J. P. Toennies, *J. Chem. Phys.*, **75**, 1794 (1981).
45. G. C. Schatz, in *Potential Energy Surfaces and Dynamics Calculations*, Ed. D. G. Truhlar, Plenum, New York, 1981, p. 287.
46. R. F. Heidner and J. V. V. Kasper, *Chem. Phys. Lett.*, **15**, 179 (1972).
47. E. B. Gordon, B. I. Ivanov, A. E. Perminov, A. E. Balalaev, A. N. Ponomarev and V. V. Filatov, *Chem. Phys. Lett.*, **58**, 425 (1978).
48. P. N. Clough, M. Kneba, U. Wellhausen and J. Wolfrum, *Discuss. Faraday Soc.*, **67**, 223 (1979).
49. W. Mayer, personal communication cited in reference 56.
50. K. T. Tang and J. P. Toennies, *J. Chem. Phys.*, **68**, 5501 (1978).
51. A. J. C. Varandas, unpublished work.
52. W. Meyer, P. C. Hariharan and W. Kutzelnigg, *J. Chem. Phys.*, **75**, 1880 (1980).
53. A. J. C. Varandas and M. C. A. Gomes, *Mol. Phys.*, **45**, 351 (1982).
54. W. Welz, *Dissertation*, University of Bonn, 1976.
55. N. Hishinuma, *J. Phys. Soc. Jpn.*, **41**, 1733 (1976).
56. F. Torello and M. G. Dondi, *J. Chem. Phys.*, **70**, 1564 (1979).
57. N. Hishinuma, *Chem. Phys. Lett.*, **70**, 361 (1981).
58. N. Hishinuma, *J. Chem. Phys.*, **75**, 4960 (1981).
59. P. Gray, *Abstracts of the 7th International Symposium on Gas Kinetics, Gottingen, 1982.*
60. J. Tennyson, *Chem. Phys. Lett.*, **86**, 181 (1982).

Chapter 9

Many-valued surfaces

9.1 Introduction

We have mentioned in several places in this book that there are many potential energy surfaces which cannot be represented for all molecular configurations by single-valued functions. The topographical features which are associated with a many-valued description are that two or more surfaces cross in some configurations but show an avoided crossing in others. The H_3 surface, described in Chapter 8, is the best-known example of a two-valued surface but, as we shall see, there are many others amongst the ground-state surfaces of triatomic systems.

The most common situation in which surface crossing occurs is when the ground state of a stable molecule cannot, by the Wigner–Witmer rules, dissociate to products which are in their ground states. We can illustrate this with the water molecule[1]. At its equilibrium configuration this has a singlet ground state of A_1 symmetry in the C_{2v} point group. Because the ground state of H_2 is a singlet and that of the oxygen atom is a triplet, it is not possible for the ground state of water to dissociate to H_2 plus O both in their ground states; this is only possible for the lowest triplet state of water. Dissociation of the singlet ground state must either give the oxygen atom in an excited singlet state or the hydrogen molecule in an excited triplet state* as shown in the following scheme:

$$H_2O(\tilde{X}, {}^1A_1) \rightarrow \begin{cases} H_2(X, {}^1\Sigma_g^+) + O({}^1D) & (a) \\ H_2(b, {}^3\Sigma_u^+) + O({}^3P) & (b) \end{cases} \qquad 9.1$$

At first sight it appears that the dissociation of H_2O, into H plus OH, poses no problems. The ground states of both H and OH are doublets and, by the Wigner–Witmer rules, these can correlate with the ground singlet state of H_2O. However, the rules governing the spatial symmetry of wave functions forbid this correlation for linear configurations of the molecule. If the bond angle of H_2O is increased to 180° with the bond lengths maintained at their equilibrium value, then the ground state becomes ${}^1\Sigma_g^+$ in $D_{\infty h}$, or ${}^1\Sigma^+$ in $C_{\infty v}$. Removing a ground-state hydrogen atom from this species along a $C_{\infty v}$ path cannot, by the conservation of angular momentum, leave the OH fragment in a ${}^2\Pi$ state. Such a dissociation

* The triplet state that dissociates to ground-state atoms is designated *b*, not *a*, as might be expected.

must lead to the lowest $^2\Sigma^+$ state of OH, which is OH(A, $^2\Sigma^+$). This dissociation of H_2O is therefore represented by the following scheme:

$$H_2O(\tilde{X}, {}^1A_1) \rightarrow \begin{cases} OH(A, {}^2\Sigma^+) + H({}^2S) & (a) \\ OH(X, {}^2\Pi) + H({}^2S) & (b) \end{cases} \tag{9.2}$$

where (a) is the required limit for linear configurations but (b) is allowed for non-linear configurations.

If one now examines the dissociation of the OH diatomic states which occur in (9.2) then one finds that the ground-state OH(X, $^2\Pi$) dissociates to ground-state atoms, H(^{2}S) plus O(^{3}P), whereas the excited state OH(A, $^2\Sigma^+$) gives H(^{2}S) plus O(^{1}D).* It follows that channels (a) of (9.1) and (9.2) are connected, and likewise channels (b). There are in fact two singlet surfaces of H_2O: one (a) which we identify with the $^2\Sigma^+$ state of OH and the ^{1}D state of O and the other (b), which we identify with the $^2\Pi$ state of OH and the ^{3}P state of O. However, we cannot identify the ground singlet state of H_2O uniquely with either (a) or (b) because one can produce either (a) or (b) dissociation limits according to the path taken to reach the dissociation limit. We have already shown that this is true for scheme (9.2) and we now look into (9.1) in more detail.

The O(^{1}D) state is 1.958 eV above the ground O(^{3}P) state but the $H_2(b, {}^3\Sigma_u^+)$–$H_2(X, {}^1\Sigma_g^+)$ energy difference depends on the H–H separation. As both states dissociate to give two ground-state hydrogen atoms, the two states are degenerate at infinite separations. From accurate representations of the two potentials of H_2 it can be shown that the $^3\Sigma_u^+ - {}^1\Sigma_g^+$ energy difference is less than 1.958 eV for H–H separations greater than 1.67 Å and it is greater than this value for separations less than 1.67 Å. Thus the ground-state potential of H_2O correlates, according to scheme (9.1), with channel (a) if the H–H separation is less than 1.67 Å, but with channel (b) if it is greater than this value.

In the next section we shall show how a two-valued potential function for the ground state of H_2O can be determined. Before commencing that task we give another example of a many-valued surface that has even greater complexity than the H_2O surface.

The ground state of the equilibrium configuration of OCS is $^1\Sigma^+$ in the point group $C_{\infty v}$. The ground states of the diatomics CO and CS are $^1\Sigma^+$ and that of OS is $^3\Sigma^-$; the ground states of all three atoms are ^{3}P. The only atom–diatom dissociation that can give two ground-state products is

$$OCS(\tilde{X}, {}^1\Sigma^+) \rightarrow OS(X, {}^3\Sigma^-) + C({}^3P) \quad (a) \tag{9.3}$$

As the C(^{3}P) state arises from the electron configuration 2p^2, its Σ component in an axial field is of Σ^- symmetry and the product of two Σ^- wave functions gives the Σ^+ symmetry of OCS as required.

For the other two atom–diatom dissociation limits one of the fragments must be in an excited state as shown in the following scheme:

* H(^{2}S) and O(^{3}P) can also combine to give a Σ state of OH but it has Σ^- and not Σ^+ symmetry.

$$\mathrm{OCS}(\tilde{X}, {}^1\Sigma^+) \rightarrow \begin{cases} \left\{\begin{array}{ll} \mathrm{CO}(a, {}^3\Pi) + \mathrm{S}({}^3\mathrm{P}) & (a) \\ \mathrm{CO}(X, {}^1\Sigma^+) + \mathrm{S}({}^1\mathrm{D}) & (b) \end{array}\right. \\ \left\{\begin{array}{ll} \mathrm{CS}(a, {}^3\Pi) + \mathrm{O}({}^3\mathrm{P}) & (a) \\ \mathrm{CS}(X, {}^1\Sigma^+) + \mathrm{O}({}^1\mathrm{D}) & (c) \end{array}\right. \end{cases} \qquad 9.4$$

The first excited triplet states of both CO and CS are $^3\Pi$ and these dissociate to ground-state atoms, as does the ground state of OS. It follows that there are three dissociation channels that can be identified with the symbol (*a*), and have a common diabatic state as indicated in the above schemes. The ground states of CO and CS also dissociate to ground-state atoms, hence scheme (*b*) would give C(^{3}P), O(^{3}P) and S(^{1}D) whereas scheme (*c*) gives C(^{3}P), O(^{1}D) and S(^{3}P). These two schemes cannot therefore be associated with the same diabatic state and it is necessary to invoke a further excited triplet state of CO which dissociates to C(^{3}P) plus O(^{1}D) and which will be produced on dissociation of the diabatic state (*c*). Likewise, there must be another triplet state of CS which dissociates to C(^{3}P) plus S(^{1}D) and which is to be associated with diabatic state (*b*).

For a complete treatment of the ground-state surface of OCS one therefore needs a three-valued representation of the surface based on the diabatic states (*a*), (*b*) and (*c*) defined above. At the present state of knowledge about the electronic states of this molecule it is unlikely that an accurate three-valued representation of this system will be obtained for some time and there is therefore a need for approximate single-valued representations of such surfaces in order that some features of the dynamics can be studied. The various methods of approaching this type of problem will be discussed in Section 9.4.

9.2 A two-valued function for the ground state of H_2O

Following the account given in the previous section on the asymptotic limits of the H_2O surface, we write the two-valued representation of the surface as the eigenvalues of the 2×2 matrix

$$\begin{pmatrix} V_{aa} & V_{ab} \\ V_{ba} & V_{bb} \end{pmatrix} \qquad 9.5$$

Because the eigenvalues of a matrix are unchanged by an orthogonal transformation there is no unique matrix representation of the H_2O surface. We can, however, provide an additional constraint which makes the matrix unique by associating the subscripts (*a*) and (*b*) with the channels (*a*) and (*b*) defined in schemes (9.1) and (9.2), and writing the diagonal elements as a many-body expansion in the same way as for a single-valued surface

$$\begin{aligned} V_{aa} &= V_{\mathrm{O}}^{(1)}({}^1\mathrm{D}) + V_{\mathrm{OH}}^{(2)}({}^2\Sigma^+, R_1) + V_{\mathrm{OH}}^{(2)}({}^2\Sigma^+, R_2) + V_{\mathrm{HH}}^{(2)}({}^1\Sigma_g^+, R_3) + V_{aa}^{(3)} \\ V_{bb} &= V_{\mathrm{OH}}^{(2)}({}^2\Pi, R_1) + V_{\mathrm{OH}}^{(2)}({}^2\Pi, R_2) + V_{\mathrm{HH}}^{(2)}({}^3\Sigma_u^+, R_3) + V_{bb}^{(3)} \end{aligned} \qquad 9.6$$

The off-diagonal term in (9.5) must go to zero at the dissociation limits, where the diabatic states give the required potential energies, and it must also go to zero for linear configurations, where there is no Σ–Π mixing. We therefore take the off-diagonal element to be a pure three-body term of the form

$$V_{ab} = V_{ba} = V_{ab}^{(3)} \sin(\hat{H}O\hat{H}) \qquad 9.7$$

The functional forms of the two- and three-body terms that occur in these expressions are chosen to be the same as similar terms for a single-valued surface and the task is reduced to finding the parameters of these functions by fitting the eigenvalues of the matrix to known properties of the potential.

There are three sources of information that can be used to determine the water potential. The first of these is spectroscopic data for the ground-state molecule. A quartic force field has been determined by Hoy *et al.*[2], and in so far as it assumes that only some off-diagonal quartic terms are zero (with good justification), it is considered to be very accurate. The second source is kinetic data on the reactions

$$O(^1D) + H_2(^1\Sigma_g^+) \rightarrow OH(^2\Pi) + H(^2S) \qquad 9.8$$

and

$$D(^2S) + OH(^2\Pi) \rightarrow OD(^2\Pi) + H(^2S) \qquad 9.9$$

As there is no activation energy for either of these reactions, there will be no barriers in the reactant channels in either case. A quasi-classical trajectory study of both systems should provide an accurate test for the surface[3].

By far the most detailed knowledge of the H_2O potential surfaces comes from *ab initio* calculations, which have been extensive, particularly in relation to the region around the equilibrium position and the reaction coordinate for (9.8). It is not easy, however, to bring together calculations made at different levels of accuracy on different parts of the surface and before a satisfactory two-valued surface for H_2O could be derived it was necessary to make further calculations covering the whole surface for the first two singlet states, using wave functions which could provide approximately the same accuracy at all points. The numerical details of these calculations can be found in the original paper[4]. The work provided approximately 150 points on the $(\tilde{X}, {}^1A')$ and $(\tilde{B}, {}^1A')$ surfaces as a basis for a non-linear least-squares optimization of the elements of the 2×2 matrix through its eigenvalues

$$V_{B/X} = \tfrac{1}{2}\{V_{aa} + V_{bb} \pm [(V_{aa} - V_{bb})^2 + 4V_{ab}^2]^{1/2}\} \qquad 9.10$$

A multidimensional non-linear optimization can generally only be achieved if one has a starting point that is not too far removed from the required optimum; without this there may be either no convergence, or convergence to a false minimum. For that reason the optimization of the H_2O potential functions was tackled in two stages. The first was to take the data points for linear configurations (where $V_{ab} = 0$) and for configurations where the two ${}^1A'$ surfaces are far apart (when V_{ab} can be neglected), and to optimize V_{aa} and V_{bb} separately to points on the two surfaces V_B and V_X. The separation of the two surfaces $|V_{aa} - V_{bb}|$ will then be less than $|V_B - V_X|$ for those regions of the surface where V_{ab} cannot be ignored, and in such regions an approximation to V_{ab} can be obtained from the separation between the two surfaces in these regions. From equation (9.10) we obtain

$$|V_{ab}| = \tfrac{1}{2}\{(V_B - V_X)^2 - (V_{aa} - V_{bb})^2\}^{1/2} \qquad 9.11$$

The second stage was a direct least-squares optimization of the eigenvalues (9.10) to the data points with the first stage functions as starting point. This gave a final root mean square error of 0.14 eV to about 150 points on the surface with a maximum error of 0.5 eV. The whole surface was then explored in regions where data were lacking to see that the functions gave no unphysical behaviour. The parameters of the final potential are given later in Table 9.2 (ref. 4).

9.3 Empirical corrections to the H_2O surface

To achieve a consistent treatment of the *ab initio* points on the H_2O surface at the asymptotic limits and in the interaction region it was necessary to take the one- and two-body terms that appear in the diagonal elements of the matrix (9.6) from calculated energies rather than the known empirical energies. However, *ab initio* calculations of this level of complexity are imperfect when tested against experimental data, as can be seen by comparing the calculated and observed dissociation limits in Table 9.1. The calculations underestimate all dissociation

Table 9.1 Comparison of experimental and calculated dissociation energies (D_e/eV) for the ground-state surface of H_2O.

Dissociation fragments	Experimental[5]	Calculated[4]
$O(^1D) + H_2(^1\Sigma_g^+)$	7.28	5.27
$H(^2S) + OH(^2\Pi)$	5.44	4.27
$H(^2S) + OH(^2\Sigma)$	9.52	8.50
$O(^3P) + 2H(^2S)$	10.07	7.68
$O(^1D) + 2H(^2S)$	12.03	9.88

energies, a deficiency which can be attributed in this case mainly to the omission of polarization functions from the basis. The contribution of polarization functions to the molecular orbitals of a system generally increases as the symmetry decreases. Thus they will contribute more to the molecular orbitals of a triatomic than to a diatomic molecule, and they will not contribute at all (by definition) to the orbitals of an atom.

It is a serious defect for a potential energy surface to have incorrect asymptotic limits because it will be incapable of reproducing the correct energy release for chemical reactions on that surface. However, the procedure that was described in the previous section for constructing the H_2O potential is very easily modified to give a surface with the correct asymptotic limits; all that is required is to replace the *ab initio* one- and two-body terms in the potential by empirical terms. This leaves only the three-body terms as *ab initio*, but as these are zero in the asymptotic limits all dissociation limits will have experimentally determined energies.

Such an empirical modification does not, of course, mean that the whole potential has empirical accuracy, but there is no doubt that the potential is improved overall by this procedure. For example, it can be seen from Table 9.1 that without the empirical correction the energy of the H_2O equilibrium configuration is -7.68 eV, relative to the energy of ground-state atoms as zero. This compares poorly with the experimental value of -10.07 eV. However, after making the correction, this energy is greatly improved to the value -10.33 eV. The implication of this result, if it can be shown to be general, is extremely important. It is that basis set deficiencies are of decreasing importance as the *n*-body expansion is ascended. This raises the possibility of making calculations with modest basis sets and improving them considerably by empirical corrections to the one- and two-body terms. This is not a new principle in quantum chemistry for it was embodied in the method of 'atoms in molecules' introduced by Moffitt[6] and is implied in the 'diatomics in molecules' method developed by Ellison[7] which was discussed in Section 3.3.

The parameters of the empirically corrected matrix elements of the two-valued H_2O surface are given in Table 9.2, and Figures 9.1–9.11 illustrate features of the surface. The first of these figures shows the intersection between the Σ and Π surfaces in linear configurations. The *ab initio* calculations show that the Π state is repulsive for the collinear approach of H to OH and hence the isotope exchange reaction (9.9) would not normally proceed along this path. However, it can be seen from Figure 9.2 that one needs only a small displacement from collinearity for the barrier to the reaction to disappear.

Figures 9.3–9.6 give a comparison of the diabatic and adiabatic surfaces for C_{2v} structures. The dotted lines in the first two figures show the intersection of the two diabatic surfaces and as one of these passes close to the equilibrium configuration of the molecule it is clear that in this region the adiabatic potentials are approximately an equal mixture of the two diabatic potentials. The upper of the two adiabatic potentials, which is associated with the $(\tilde{B}, {}^1A_1)$ excited state of the molecule, has a minimum in a linear configuration, in agreement with spectroscopic evidence. The secondary minimum which occurs with the oxygen atom about 2 Å away from a normal H_2 molecule is probably not physically real and it arises from the fact that there are other ${}^1A'$ states in this region other than those we have considered which bring additional complications to this excited state surface but not to the ground-state surface. Figures 9.7–9.10 give representations of the diabatic and adiabatic surfaces for the motion of one hydrogen atom around an OH molecule with its bond length fixed at its ground state equilibrium distance.

Finally, Figure 9.11 shows the lowest adiabatic surface for the approach of $O({}^1D)$ to $H_2({}^1\Sigma_g^+)$ with the diatomic bond length fixed at its equilibrium value. This shows, together with Figure 9.5, that there is no barrier for any direction of approach.

Table 9.2 Parameters of the two-valued H_2O potential (9.5) whose elements are defined by equations (9.6) and (9.7). The two-body terms are ER functions (1.19), and the three-body terms are the product of (6.5) and (6.6). By symmetry $C_1 = C_2$, etc.

One-body term: $V_O^{(1)} = 1.958$ eV

Two-body terms:

Species	D_e/eV	R_e/Å	a_1/Å^{-1}	a_2/Å^{-2}	a_3/Å^{-3}	γ/Å^{-1}
$OH(^2\Sigma^+)$	2.5034	1.0121	3.7533	0.0	0.0	3.7533
$OH(^2\Pi)$	4.6316	0.9700	4.5347	5.0305	3.9254	4.5347
$H_2(^3\Sigma_u^-)$	−5.8568	0.7417	1.4782	1.7802	0.0	3.9022
$H_2(^1\Sigma_g^+)$	4.7500	0.7412	3.9525	4.0300	3.5372	3.9525

Three-body terms:

Parameter	$V_{aa}^{(3)}$	$V_{bb}^{(3)}$	$V_{ab}^{(3)}$
V^0/eV	0.0415	1.4965	5.1015
R_1^0(OH)/Å	1.4520	0.6800	0.8098
R_3^0(HH)	2.4140	1.3393	0.6640
γ_1(OH)/Å^{-1}	3.0953	1.6469	1.6422
γ_3(HH)	−1.8183	0.5297	−6.2544
C_1/Å^{-1}	51.5130	1.7062	1.1740
C_3	−37.2482	16.9766	−1.7485
C_{11}/Å^{-2}	−71.7768	−15.1013	0.3868
C_{33}	−21.5248	29.5816	2.0734
C_{12}	−94.6933	42.7626	2.9627
C_{13}	91.6818	−43.9501	−1.9907
C_{111}/Å^{-3}	94.5128	12.0197	−0.8544
C_{333}	−126.4926	8.6860	−1.0265
C_{112}	237.1390	−17.3648	−0.8402
C_{113}	−192.4370	22.1420	0.4452
C_{133}	238.0566	−28.5625	1.1476
C_{123}	−391.2772	49.8140	−1.6264
C_{1111}/Å^{-4}	−14.2493	−3.6062	−0.0294
C_{3333}	19.9178	−1.7078	0.4344
C_{1112}	−16.5365	−2.4602	−3.1701
C_{1122}	−10.0277	6.3071	7.5300
C_{1113}	−22.0173	2.3598	0.2732
C_{1133}	7.3286	−2.4987	0.6041
C_{1333}	−3.7564	3.0869	−1.0816
C_{1123}	−13.8227	−2.3485	−1.3691
C_{1233}	−20.4843	0.8883	2.5884

9.4 Single-valued representations of the H_2O surface

The two-valued representation of the H_2O potential that has been described in the last two sections was developed to satisfy the asymptotic requirements of the

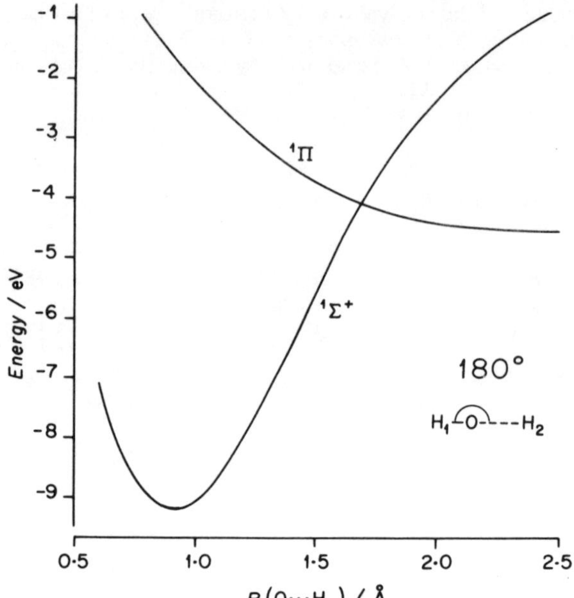

Figure 9.1 The intersection of $^1\Sigma^+$ and $^1\Pi$ surfaces for linear configurations $R(H_1\text{–}O) = 0.92$ Å

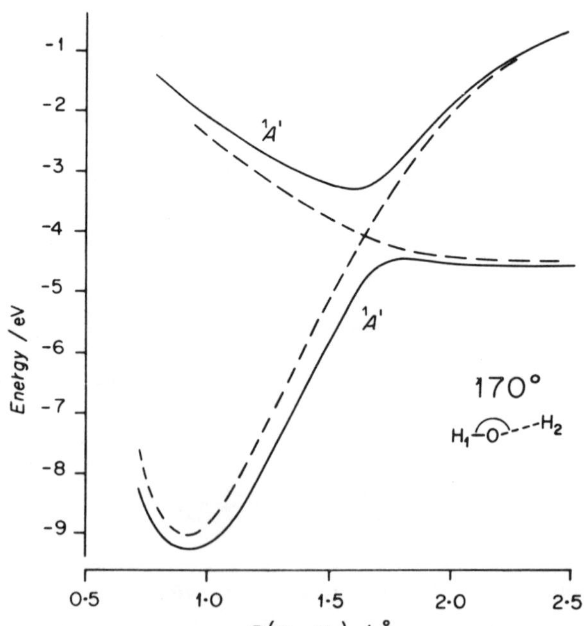

Figure 9.2 The avoided intersection of the $^1A'$ surfaces for small displacements from linear configurations. The dashed lines are the diabatic potentials

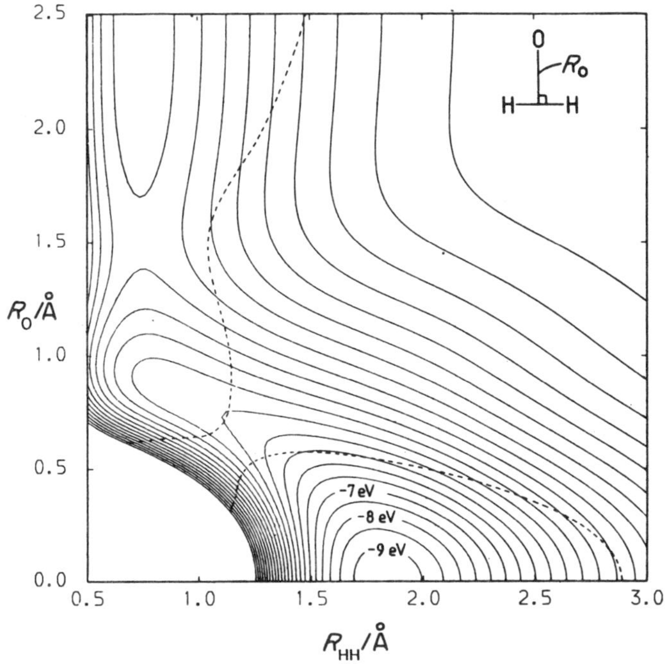

Figure 9.3 Contours of the Σ–^{1}D diabatic potential V_{aa} [equation (9.6)] for C_{2v} configurations. The dashed curves are the intersection with V_{bb}

potential which are summarized in schemes (9.1) and (9.2). However, this level of complication should not be introduced unless it is shown to be necessary for the dynamical problems to which the surface is to be applied. For example, scheme (9.1) shows that dissociation of the ground state of H_2O can, in principle, lead to $H_2(b, {}^3\Sigma_u^+) + O({}^3P)$ [channel (*a*)], but as the triplet state of H_2 is repulsive this in effect means dissociation to $O({}^3P)$ plus two hydrogen atoms. However, such a process requires a very high energy, one that is greatly in excess of that needed to give $H_2(X, {}^1\Sigma_g^+) + O({}^1D)$ [channel (*b*)]. Channel (*a*) can therefore be ignored except at exceptionally high energies. In a similar vein, scheme (9.2) includes the possibility of dissociation to $H({}^2S) + OH(A, {}^2\Sigma^+)$, yet we have seen that this will only occur for classical trajectories that proceed exactly along collinear configurations and these have infinitesimal probability.

The point we are stressing is that for many dynamical problems it is not necessary to have an accurate representation of the potential for all nuclear configurations. Some regions may be uninteresting because they have a high energy, and others may have a low energy but be separated from the significant region by high barriers. With such restrictions in mind it may be sufficient to use a potential function which is simpler than the topographically exact potential, but is adequate to its task. In this spirit there have been several single-valued representations of the ground-state H_2O potential which have been developed and used for dynamical studies. The first of these, by Sorbie and Murrell[1], was in

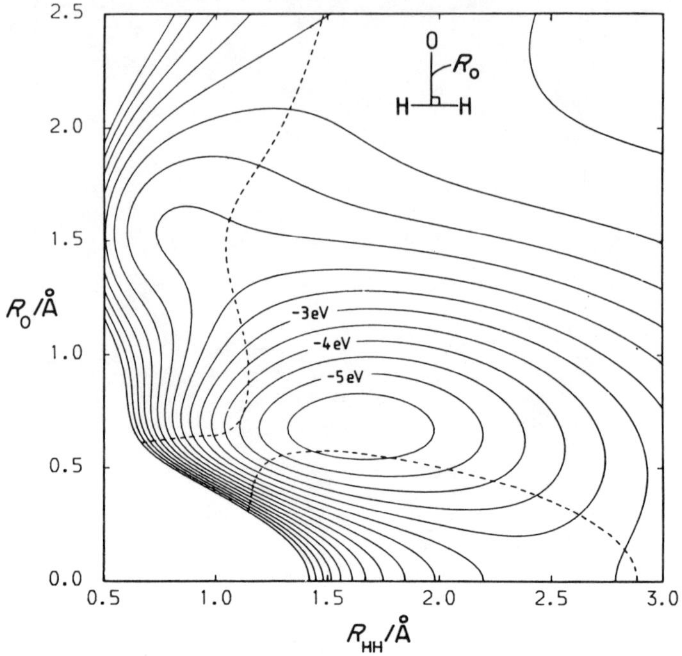

Figure 9.4 Contours of the Π–^{3}P diabatic potential V_{bb} [equation (9.6)] for C_{2v} configurations. The dashed curves are the intersection with the diabatic potential V_{aa}

fact the first paper in which the many-body expansion of the potential was developed, so it justifies some attention even though its approach has been superseded by later ideas.

The asymptotic limits contained in scheme (9.1) can be imposed on the many-body expansion of the potential by defining a two-body H–H term which follows the $^1\Sigma_g^+$ potential (raised by the oxygen ^{1}D–^{3}P energy) when channel (*a*) has the lower energy and the $^3\Sigma_u^+$ potential when channel (*b*) has the lower energy. This potential has a cusp when the two channels are equal, which occurs at $R_{HH} = R_{cusp} = 1.67$ Å, as shown in Figure 9.12. The difficulty with this approach is that the cusp in this two-body term should appear in the total potential only at the asymptotic limit in which the O and H_2 are infinitely separated; in the interaction region it is therefore necessary to introduce a similar cusp in the three-body term which should, ideally, exactly cancel the two-body cusp and leave the total potential differentiable.

Sorbie and Murrell[1] suggested a recipe for producing a satisfactory potential with a cusped two-body term. They divided the total space into two regions according to the value of R_{HH} and defined potentials $V_{in}(R_{HH} \leqslant R_{cusp})$ and $V_{out}(R_{HH} \geqslant R_{cusp})$ in the two regions. They then restricted the three-body terms in these two regions to have the same order of polynomial, P, and identical range functions, T. Thus continuity of energy across R_{cusp} requires

$$[V_{HH}^{(2)}(\text{in}) + P_{in}T]_{R_{cusp}} = [V_{HH}^{(2)}(\text{out}) + P_{out}T]_{R_{cusp}} \qquad 9.12$$

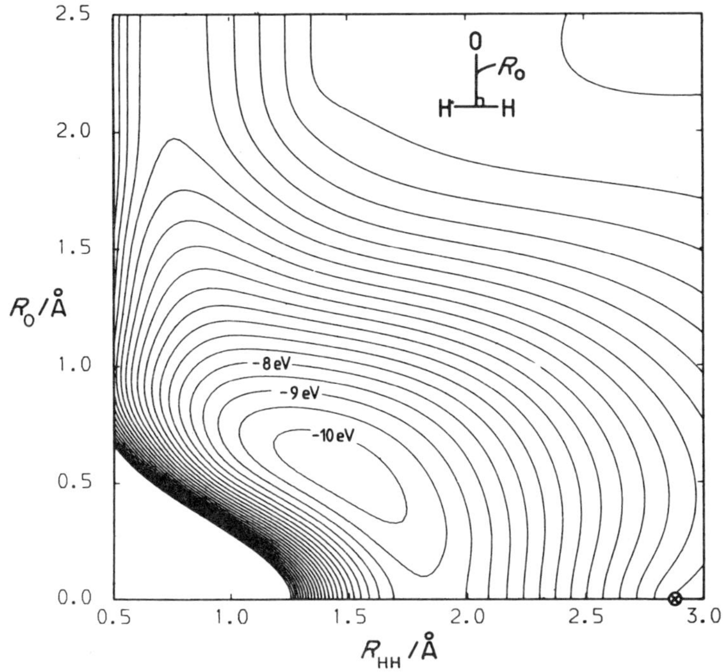

Figure 9.5 Contours of the empirically corrected adiabatic ground state surface, $H_2O(\tilde{X})$, for C_{2v} configurations. The point marked ⊗ is the Σ^+/Π conical intersection with the upper surface $H_2O(\tilde{B})$

which can be re-written as

$$(P_{out} - P_{in})_{R_{cusp}} = \{[V_{HH}^{(2)}(in) - V_{HH}^{(2)}(out)]T^{-1}\}_{R_{cusp}} \qquad 9.13$$

If a Taylor expansion of the right-hand side of equation (9.13) is made about the point $R_{OH} = R_{OH'} = R_{OH}^e$, $R_{HH} = R_{cusp}$, then coefficients of like terms on both sides of this equation can be equated to provide a series of equations connecting the coefficients of the polynomials P_{in} and P_{out}. This recipe does not provide differentiability in R_{HH} for all values of the OH distances, but it has been shown in practice to provide an acceptably smooth potential for fairly large displacements of R_{OH} from their equilibrium values.

Finally, it should be mentioned that the Sorbie–Murrell H_2O potential was determined by fitting the coefficients of P_{in} to the quartic force field of H_2O, following the procedure described in Section 7.1. Subsequent variational calculations of the vibration frequencies of H_2O (ref. 1) have shown that the potential is very accurate in the region of the equilibrium configuration. However, in other respects it was less satisfactory. The failure to satisfy the spatial correlation rules (9.2) with this potential leads to the absence of a barrier for the collinear approach of H to OH, and this would probably make the surface unsatisfactory for studying the isotope exchange reaction (9.9).

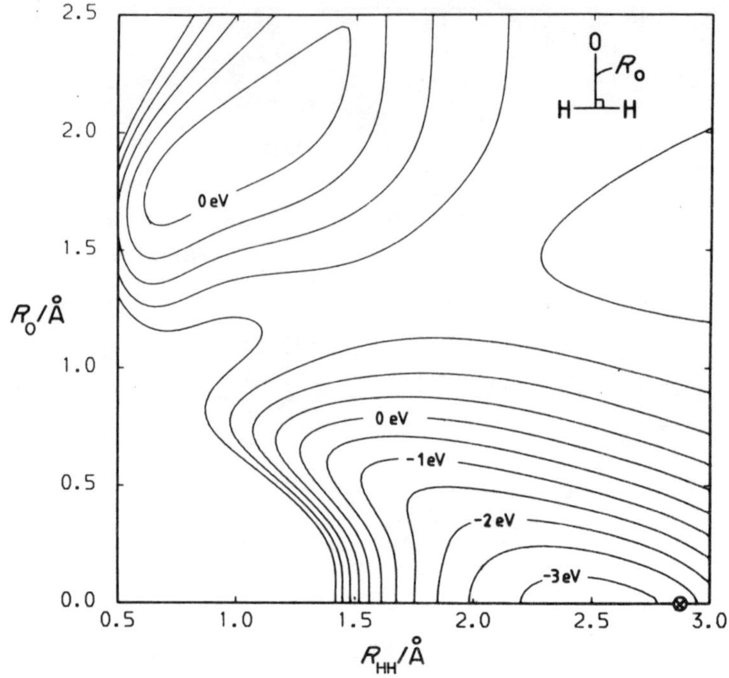

Figure 9.6 Contours of the empirically corrected adiabatic surface for the excited $\tilde{B}$ state of H_2O, for C_{2v} configurations. The point marked ⊗ is the intersection with the ground state

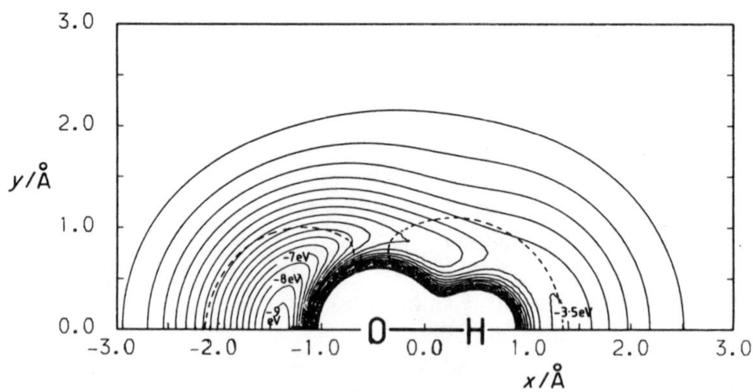

Figure 9.7 Contours of the Σ–^{1}D diabatic potential V_{aa} [equation (9.6)] for the movement of H around OH with a fixed O—H bond length of 0.97 Å. The dashed curves are the intersection with the diabatic potential V_{bb}

An alternative approach to a single-valued H_2O surface was made by Schinke and Lester[8]. Instead of using a cusped two-body term for the H–H interaction they joined the $O(^1D) + H_2(^1\Sigma_g^+)$ potential to the $2H + O(^3P)$ asymptotic limit by a monotonically increasing exponential function

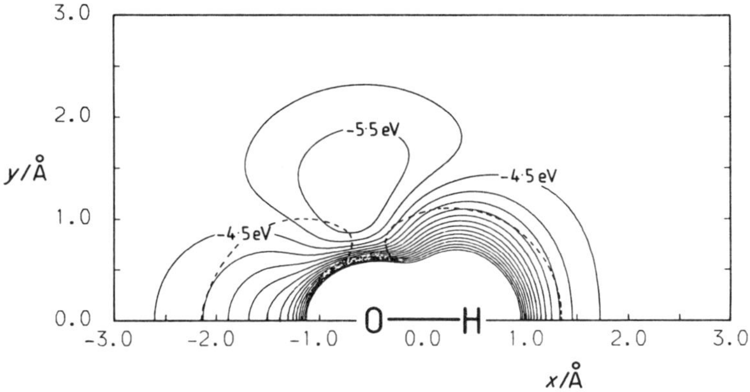

Figure 9.8 Contours of the diabatic potential V_{bb}. R(O—H) = 0.97 Å

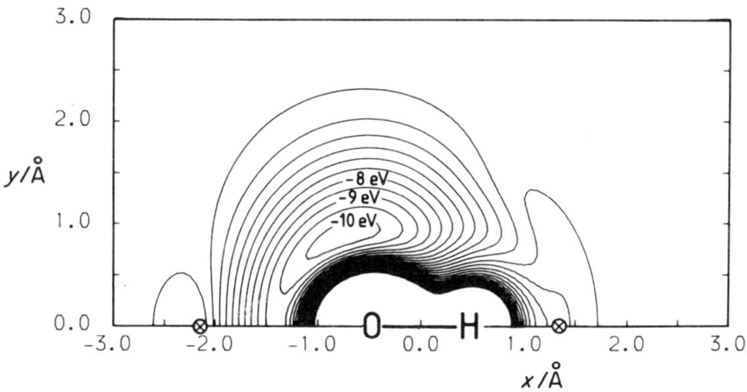

Figure 9.9 Contours of the ground-state adiabatic potential. R(O–H) = 0.97 Å. ⊗ = Σ^+–Π intersection

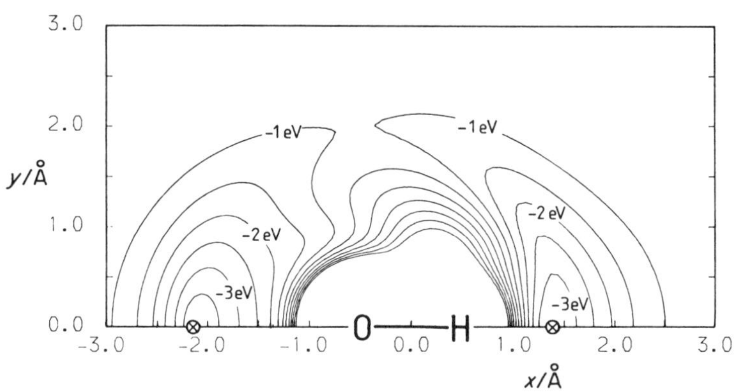

Figure 9.10 Contours of the excited state adiabatic potential. R(O–H) = 0.97 Å. ⊗ = Σ^+–Π intersection

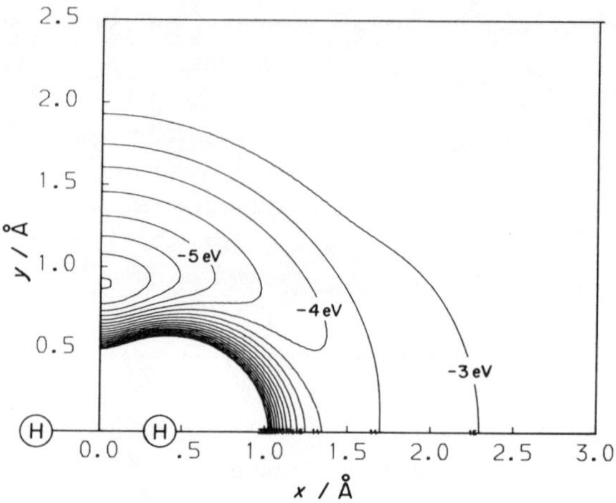

Figure 9.11 Contours of the ground-state adiabatic potential for the approach of $O(^1D)$ to $H_2(^1\Sigma_g^+)$ with R(H–H)=0.74 Å

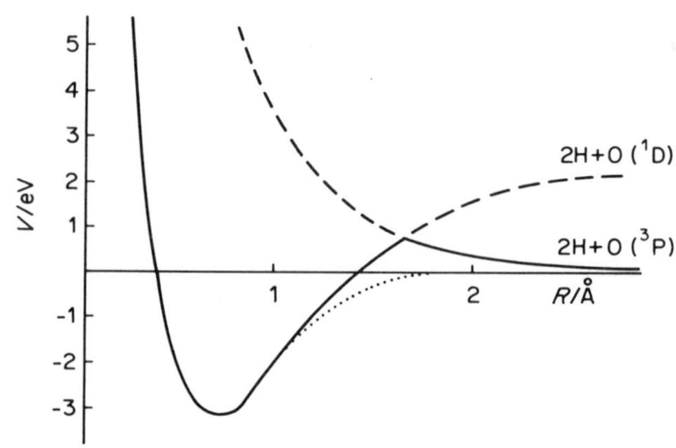

Figure 9.12 The two-body HH potentials used by Sorbie and Murrell[1] (——) and by Schinke and Lester[8] (·····), to obtain single-valued representations of the H_2O potential. R is the H–H distance, and the dashed curves show the continuation of the two branches in the Sorbie–Murrell potential along the $^1\Sigma_g^+$ and $^3\Sigma_u^+$ potentials of H_2

$$V_{HH}^{(2)} = B \exp\left[-\alpha(R - R_e)\right] \qquad 9.14$$

for $R_{HH} \geqslant 1.1$ Å. It can be seen from Figure 9.12 that this represents an appreciable deviation from the $H_2(^1\Sigma_g^+)$ potential at fairly low energies and any dynamical calculations which would involve H_2 above the $v=1$ state would not be realistic. However, for the reactions of ground-state H_2, which are much more important, the potential will be entirely adequate.

As the two-body term used by Schinke and Lester is not cusped, it may be combined with an analytic three-body term to obtain a representation of the whole surface. Schinke and Lester[8] used for this a similar function to that of Sorbie and Murrell, a product of a polynomial and a range function, fitting the parameters by a least-squares criterion to *ab initio* points on the surface.

There are clearly many other functions that provide a smooth interpolation between the two branches of the H–H two-body term, and that used by Schinke and Lester seems unnecessarily simple. In general terms we require a switching function between the two branches with the centre and range of this function depending on the coordinates of the system. The most commonly employed switching function is in one dimension, of the form

$$L(x)=\tfrac{1}{2}[1-\tanh(\alpha x)] \qquad 9.15$$

which has the limits unity as $x\rightarrow-\infty$ and zero as $x\rightarrow\infty$. It follows that the function

$$f(x)L(x)+g(x)L(-x) \qquad 9.16$$

approaches $f(x)$ as $x\rightarrow-\infty$ and $g(x)$ as $x\rightarrow\infty$ and the rate at which the function switches from $f(x)$ to $g(x)$, through $x=0$, increases as α increases.

As the switching function (9.15) has the same form as the range functions employed for the three-body term (3.25), it is natural that it should be used in the scheme which we have developed in this book. The task before us is to decide the relationships between the variables x and the coordinates of the system, and then to determine the values of the parameters α. One approach, which would be consistent with the treatments of Sorbie and Murrell, and Schinke and Lester for the H_2O surface, would be to take x to be the coordinate of the two-body potential to which the switching function is attached; in the H_2O case the H–H distance. However, this is unnecessarily restrictive as in more complex systems we may wish to have different asymptotic limits in several different regions of space. Our preferred approach, which we shall describe[9], has this flexibility. The method imposes the condition that in a chemically distinguishable asymptotic region of the surface there is only one possible set of electronic states for the fragments. Indeed, without this restriction there is no possibility of using a single-value representation of the surface, for two-valuedness is essentially an expression of the possibility of two distinguishable sets of states in any asymptotic region. We illustrate our approach by the example of the ground state of H_2O and in Section 12.4 show its extension to the tetra-atomic system H_2O_2.

In contrast to the two complete dissociation schemes for the H_2O surface (9.1) and (9.2), we allow only the following possibilities

$$H_2O(\tilde{X},{}^1A_1)\rightarrow \begin{array}{l} H_2(X,{}^1\Sigma_g^+)+O({}^1D) \\ OH(X,{}^2\Pi)+H({}^2S) \end{array} \qquad 9.17$$

A comparison with (9.1) shows that we have not allowed for H_2 to be removed in its triplet state. However, as the triplet state is repulsive this means that our allowed channel is below the three-atom dissociation limit $2H({}^2S)+O({}^3P)$ and hence for most practical dynamics on the ground-state H_2O surface we have the only allowed channel of interest. Also, we have not allowed for the production of

$OH(A, {}^2\Sigma^+)$, but as this requires at least 4.1 eV more energy than the production of ground-state OH, this also is generally of no interest.

It can be seen that the only requirement to satisfy (9.17) is to switch off the one-body energy for $O({}^1D)$ in the asymptotic regions corresponding to the limits OH+H. If we now consider the variable

$$x = n\rho_3 - \rho_1 - \rho_2 \qquad 9.18$$

where ρ_3 is the HH distance and ρ_1 and ρ_2 the OH distances, all relative to a reference structure, $\rho_i = R_i - R_i^0$, then we see that x takes the limit $-\infty$ for dissociation into H_2+O and $+\infty$ for dissociation to OH+H provided that $n \geqslant 2$. For the three-atom limit 2H+O, equation (9.18) has no single limit for all space, although for large n, in most space, it takes the limit $+\infty$.* As the $O({}^1D)$ energy should be switched off in the three-atom limit, in order to obtain the lowest energy products, a high value of n appears to have some advantages, and we have chosen $n=3$ for our analysis. We therefore write the switching function for the one-body energy $V_O^{(1)}$ as

$$f = \tfrac{1}{2}[1 - \tanh(3\rho_3 - \rho_1 - \rho_2)\alpha/2] \qquad 9.19$$

and the single-valued representation of the H_2O surface as

$$V_{HHO} = V_O^{(1)} f(R_1, R_2, R_3) + V_{OH}^{(2)}(R_1) + V_{OH}^{(2)}(R_2) + V_{HH}^{(2)}(R_3) + V_{HHO}^{(3)}(R_1, R_2, R_3) \qquad 9.20$$

where the two-body energies are those appropriate to the diatomics appearing in (9.17), and $V_{HHO}^{(3)}$ is a three-body term of the form (3.26).

The parameters in the potential (9.20) which have to be optimized are the range parameters α and γ_i, and the coefficients of the three-body term. For any values of the range parameters it is possible to obtain coefficients of a quadratic polynomial which reproduces the energy, geometry and harmonic force field of the equilibrium molecule. Surfaces thus obtained can then be scanned over the whole space to see if they have any unphysical behaviour, e.g. other minima or maxima for which there is no physical evidence. This was done[9] for variations of α, $\gamma_1 = \gamma_2$ and γ_3 in the range 1–4 Å^{-1} and the most satisfactory potential was judged to be that with $\alpha = 1.5$ Å^{-1}, $\gamma_1 = \gamma_2 = 2.6$ Å^{-1} and $\gamma_3 = 1.5$ Å^{-1}.

The next step was to calculate, by the variation method, the vibration frequencies of this best surface and to compare them with experimental frequencies. The differences between observed and calculated frequencies could then be minimized by extending the three-body polynomial to a full quartic in the three variables ρ_i and optimizing these coefficients and the value of α, using the procedure described in Section 7.3. Further optimization with respect to the γ_i is also possible but was not found to lead to any improvement.

Table 9.3 gives the parameters of the best potential that we have obtained, and

* For example, in linear configurations H—H—O, we have $R_1 = R_2 + R_3$, hence equation (9.18) becomes $(n-1)R_3 - 2R_2(nR_3^0 - R_1^0 - R_2^0)$. As R_2 and R_3 are independent variables the asymptotic limit of equation (9.18) is undefined.

Table 9.3 Parameters of the H_2O potential whose functional form is given by equations (9.19) and (9.20). The two-body terms differ slightly from those in Table 9.2 as those given here were obtained from a later, more accurate, fit to diatomic data.

One-body terms: $V_O^{(1)} = 1.958$ eV; $\alpha = 1.9018$ Å^{-1}

Two-body terms:

Species	D_e/eV	R_e/Å	a_1/Å^{-1}	a_2/Å^{-2}	a_3/Å^{-3}
OH	4.6211	0.9696	4.507	4.884	3.795
HH	4.7472	0.7414	3.961	4.064	3.574

Three-body term:

$C_1 = C_2$/Å^{-1}	−13.488	$C_{1111} = C_{2222}$/Å^{-4}	189.824
C_3	−34.802	C_{3333}	626.320
$C_{11} = C_{22}$/Å^{-2}	65.063	$C_{1112} = C_{1222}$	502.244
C_{33}	414.907	$C_{1113} = C_{2223}$	−116.088
C_{12}	123.533	$C_{1333} = C_{2333}$	−760.070
$C_{13} = C_{23}$	−429.933	C_{1122}	−560.202
$C_{111} = C_{222}$/Å^{-3}	114.512	$C_{1133} = C_{2233}$	−255.785
C_{333}	−56.338	$C_{1123} = C_{1223}$	139.484
$C_{112} = C_{122}$	−374.139	C_{1233}	153.098
$C_{113} = C_{223}$	−187.587	V^0/eV	0.01892
$C_{133} = C_{233}$	209.543	$\gamma_1 = \gamma_2$/Å^{-1}	2.6
C_{123}	−271.062	γ_3	1.5
		R_1^0(OH)/Å	0.9572
		R_3^0(HH)	1.5139

Table 9.4 compares the calculated and observed vibration frequencies. The agreement between the two is very similar to that obtained with the Sorbie–Murrell potential.

Figures 9.13–9.15 show views of the surface and a comparison with the lower sheet of the two-valued representation can be made (Figures 9.5 and 9.9). The only significant difference is in the region of linear configurations, both H—H—O and H—O—H, where the Σ–Π conical intersections should exist. These cannot, of course, be reproduced by a single-valued representation.

9.5 Renner–Teller potentials

Before leaving this chapter we summarize some conclusions that have been reached by Carter *et al.*[10] on the form of the potential energy functions for systems that are orbitally degenerate in linear configurations, the so-called Renner–Teller systems[11]. The question of importance is what is the dependence of the potential energy functions on the bending coordinate α.

There are three classes of Renner–Teller surfaces that must be considered. In the first the two surfaces that arise from the orbitally degenerate state (which we call the Renner–Teller pair of surfaces) do not intersect with any other electronic surface along a linear dissociation pathway. The state $C_3(\tilde{A}, {}^1\Pi_u)$ is of this type. In

Table 9.4 Comparison of observed and variationally calculated vibration wavenumbers (cm^{-1}) of the H_2O surface. Values in parentheses are not observed frequencies but our own upper bound estimates.

	H_2O		D_2O	
Level	Calc.	Obs.	Calc.	Obs.
100	3649	3657	2669	2671
010	1596	1595	1174	1178
001	3762	3756	2792	2788
200	7176	7201	5279	5292
020	3150	3151	2335	(2356)
002	7453	7445	5541	(5576)
110	5249	5235	3844	(3849)
101	7245	7250	5375	5374
011	5336	5331	3956	3956
300	10817	(10858)		
030	4641	4667		
003	11044	11032		
210	8772	8762		
120	6797	6775		
201	10596	10613		
102	11102	(11102)		
021	6874	6871	5107	5105
012	9005	9000		
111	8819	8807	6538	6533
130	8282	8274		
031	8359	8374		
013	12581	12565		
211	12161	12151		
121	10344	10329		

the second class one component of the Renner–Teller pair intersects the surface of a non-degenerate Σ state. The ground-state surface of NH_2 which is of symmetry $^2\Pi_u$ in $D_{\infty h}$ configurations shows an intersection with a $^2\Sigma_g^-$ state and is of this class. The third class is that for which both components intersect with Σ^+ or Σ^- surfaces or with other orbitally degenerate pairs. The ground-state surface of HNO is $^1\Delta$ in linear configurations and this intersects with a $^1\Pi$ surface.

An orbitally degenerate surface for a linear ABC system will, on bending, separate into A' and A'' surfaces, and separate matrix representations can be obtained for the two symmetry species. For example, for the $^1\Pi_u$ surfaces of C_3 there will be single-valued representations (1×1 matrices) for both the A' and A'' components. For NH_2 there will be a 1×1 A' matrix and a 2×2 A'' matrix. Finally, for HNO 2×2 matrices will be needed for both A' and A'' surfaces.

In zero-order the diabatic (diagonal) elements of the A' and A'' components will be identical, but the perturbations associated with bending the molecule through an angle α will be different. The differences between the A' and A'' potentials can be thought of as contributions to the energy from the orbital angular momentum.

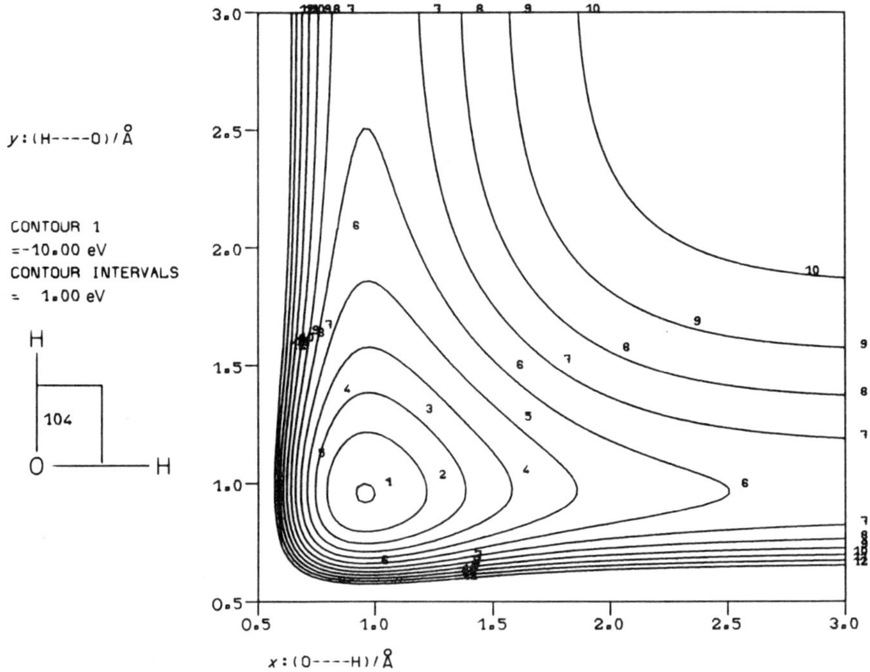

Figure 9.13 Contours of the H_2O potential given in Table 9.3

It has been shown by Carter *et al.*[10] that topographically correct potentials can be obtained by writing the potential matrix elements as

$$V'_{rr} = V^0_{rr} + \sin^n \alpha \cdot V'^{(3)}_{rr}$$

$$V'_{rs} = \sin^m \alpha \cdot V'^{(3)}_{rs} \qquad 9.21$$

for the A' matrices, and

$$V''_{rr} = V^0_{rr} - \sin^n \alpha \cdot V''^{(3)}_{rr}$$

$$V''_{rs} = \sin^m \alpha \cdot V''^{(3)}_{rs} \qquad 9.22$$

for the A'' matrices, where

$$n = \lambda_r^+ - \lambda_r^- \qquad 9.23$$

and

$$m = |\lambda_r^+ - \lambda_s^+| = |\lambda_r^- - \lambda_s^-| \qquad 9.24$$

λ_r^+ and λ_r^- are the orbital angular momentum quantum numbers for the two components of the Renner–Teller pair and λ_s^+ and λ_s^- are similarly the quantum numbers for the other surfaces (if any) that are crossed. For example, for a Δ state we have $\lambda_r^+ = 2$, $\lambda_r^- = -2$, and hence $n = 4$. If this crosses a Π state (as for HNO), then the off-diagonal element has a $\sin \alpha$ functionality. $V^{(3)}$ in each of these expressions is a three-body term.

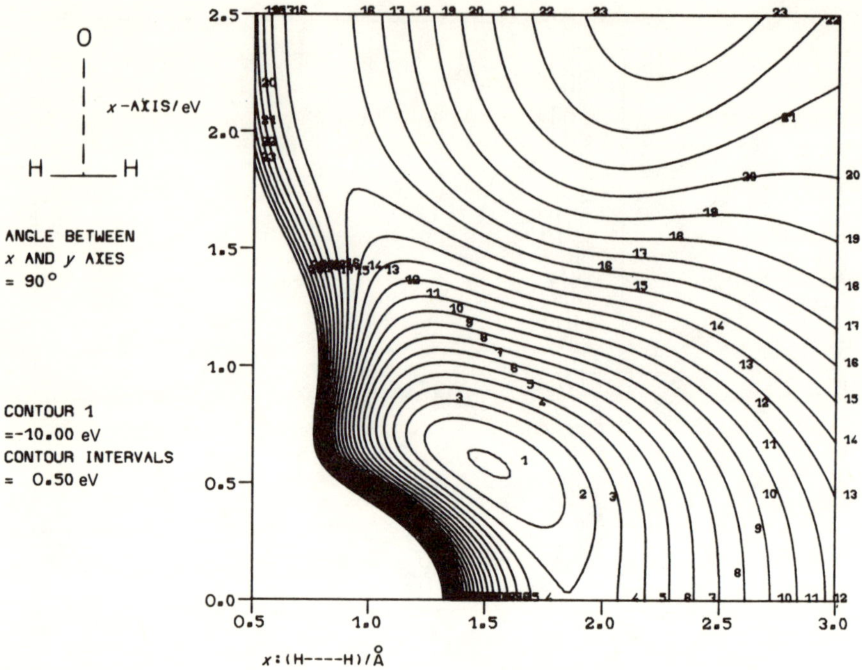

Figure 9.14 Contours of the H_2O potential given in Table 9.3. This figure should be compared with the corresponding Figure 9.5 for the two-valued potential

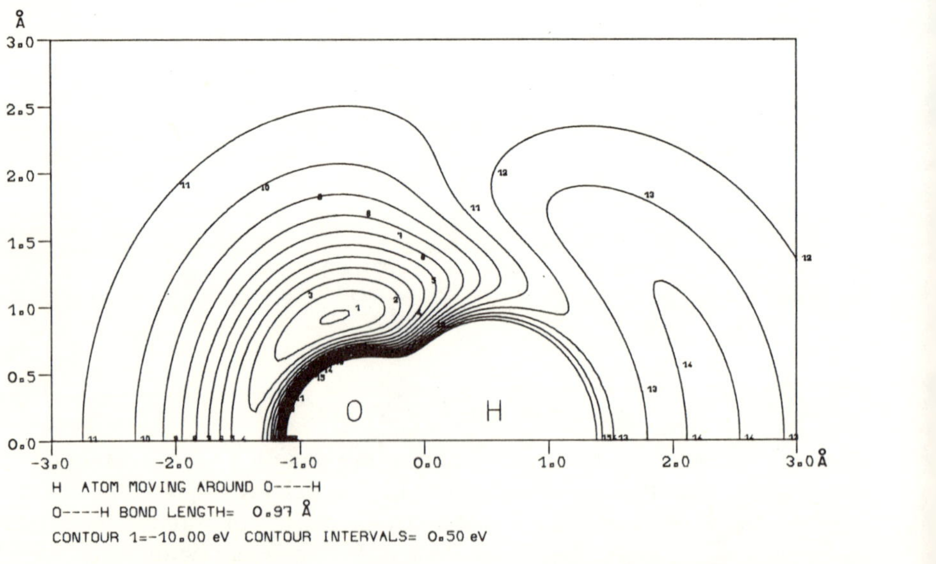

Figure 9.15 Contours of the H_2O single-valued potential of Table 9.3 for the movement of H around OH. This figure should be compared with Figure 9.9

From the diagonal elements of these matrices it can be seen that the two surfaces of a linear Π state separate as α^2 for small angles and Δ states separate as α^4 and this is the observed behaviour. The intersection of a Σ state and a Π state, for example, gives a conical intersection with an off-diagonal element depending on sin α. The two-valued surface of H_2O discussed in Section 9.2 conforms to this rule.

References

1. K. S. Sorbie and J. N. Murrell, *Mol. Phys.*, **29**, 1387 (1975).
2. A. R. Hoy, I. M. Mills and G. Strey, *Mol. Phys.*, **24**, 1265 (1972).
3. L. Dunne and J. N. Murrell, *Mol. Phys.*, **50**, 635 (1983).
4. J. N. Murrell, S. Carter, I. M. Mills and M. F. Guest, *Mol. Phys.*, **42**, 605 (1981).
5. *JANAF Tables of Thermochemical Data*, NBS Technical Note, 2nd ed., National Bureau of Standards, Washington, DC, 1971.
6. W. Moffitt, *Proc. R. Soc. London, Ser. A*, **210**, 245 (1951).
7. F. O. Ellison, *J. Am. Chem. Soc.*, **85**, 3540 (1963).
8. R. Schinke and W. A. Lester, *J. Chem. Phys.*, **72**, 3754 (1980).
9. S. Carter and J. N. Murrell, *J. Phys. Chem.*, in press.
10. S. Carter, I. M. Mills and R. N. Dixon, to be published.
11. G. Herzberg, *Molecular Spectra and Molecular Structure, III, Electronic Spectra of Polyatomic Molecules*, Van Nostrand, Princeton, 1966.

Chapter 10

Application of the many-body expansion to surfaces with several minima

In Section 4.2 we showed how the internal coordinates for a triatomic system could be chosen to reproduce the permutation symmetry of the surface, and in Section 4.3 we showed that even in the absence of symmetry there are optimum coordinates for a surface possessing several minima. In both cases the coordinates used, S_i, are a linear transformation of the internuclear displacement coordinates $\rho_i = R_i - R_i^0$. Most surfaces for triatomic molecules have more than one minimum, sometimes related by symmetry and in other cases not. We have already dealt with three such systems, SO_2, C_3 and HCN, and in this section we deal with three others. Our aim is, by example, to show the general method of obtaining potential functions for such surfaces.

10.1 The O_3 surface

The equilibrium geometry of the ground state of O_3 has C_{2v} symmetry, with a bond angle of 117°. By symmetry there must be three such minima on the surface all equivalent, because any one of the oxygen atoms can be the central atom. In addition, there is evidence from *ab initio* calculations for the existence of a metastable D_{3h} minimum.

In a preliminary study of the surface[1], it was shown that a potential determined by fitting the energy, geometry and force field of one C_{2v} structure without taking account of symmetry did show two other minima but these were at linear rather than C_{2v} configurations of the system. It is therefore clear that symmetry must be imposed on the potential function in order to obtain the correct topography of the surface, and this may be done by expressing the potential in the symmetry coordinates for an X_3 system (4.22). In terms of displacement coordinates ρ_i from a D_{3h} reference structure of bond length R^0 these are ($\rho_i = R_i - R^0$)

$$\begin{bmatrix} S_1 \\ S_2 \\ S_3 \end{bmatrix} = \begin{bmatrix} \sqrt{\frac{1}{3}} & \sqrt{\frac{1}{3}} & \sqrt{\frac{1}{3}} \\ 0 & \sqrt{\frac{1}{2}} & -\sqrt{\frac{1}{2}} \\ \sqrt{\frac{2}{3}} & -\sqrt{\frac{1}{6}} & -\sqrt{\frac{1}{6}} \end{bmatrix} \begin{bmatrix} \rho_1 \\ \rho_2 \\ \rho_3 \end{bmatrix} \qquad 10.1$$

As discussed earlier, the totally symmetric combinations of these S_i coordinates can all be expressed as products of the integrity basis (4.18):

$$S_1,\ S_2^2+S_3^2,\ S_3^3-3S_3S_2^2 \qquad 10.2$$

If the reference structure is defined to have the same perimeter as the C_{2v} equilibrium structure

$$R^0=\tfrac{1}{3}(R_1^0+R_2^0+R_3^0) \qquad 10.3$$

then $S_1=0$ at the C_{2v} minima, and this simplifies the relationship between the derivatives of the potential at the minima and the coefficients of the three-body term. The three C_{2v} minima have different values of the variables S_2 and S_3. If the one with $R_2=R_3$ has coordinates (0, 0, d), then the other two are at (0, $\sqrt{3}\,d/2$, $d/2$) and (0, $\sqrt{3}\,d/2$, $-d/2$).

The energy, geometry (bond length and bond angle) and the harmonic force field (f_{rr}, $f_{rr'}$, $f_{\alpha\alpha}$ and $f_{r\alpha}$) provide seven pieces of data, which is more than enough to determine the coefficients of a quadratic polynomial but not sufficient to determine all the coefficients of a quartic polynomial. To see how the data and coefficients are related we start with the full flexibility of the quartic polynomial and write the three-body term as

$$\begin{aligned} V^{(3)}=V^0[&1+C_1S_1+C_2S_1^2+C_3(S_2^2+S_3^2)+C_4S_1^3+C_5S_1(S_2^2+S_3^2) \\ &+C_6S_3(S_3^2-3S_2^2)+C_7S_1^4+C_8S_1^2(S_2^2+S_3^2)+C_9S_1S_3(S_3^2-3S_2^2) \\ &+C_{10}(S_2^2+S_3^2)^2](1-\tanh\gamma S_1/2) \end{aligned} \qquad 10.4$$

Table 10.1 lists the derivatives with respect to symmetry coordinates of $V^{(3)}$ evaluated at (0, 0, d). Each of these derivatives can be determined from the data referred to above.

From the first four equations in Table 10.1 (a)–(d) we can determine V^0, C_3, C_6 and C_{10}, which are independent of γ. From equations (e) and (f) we can determine two of C_1, C_5 and C_9 and from (g), one of C_2 and C_8. On the principle that we select for preference the polynomial terms of lowest order, then we delete the terms in C_8 and C_9 from the three-body term given in equation (10.4). Such terms could only be included if fresh data were added, for example, higher terms in the force field.

We shall now look at the three-body polynomial from another point of view and pose the following question. If the coefficients were to be determined from a set of *ab initio* points on the surface, then at what points on the surface should calculations be made? If we start with a series of D_{3h} structures with different bond lengths then these can be used to determine the coefficients of terms independent of S_2 and S_3 (which are both zero in D_{3h}). If we now proceed to

Table 10.1 Derivatives of the three-body term (10.4) in terms of symmetry coordinates, $Y_{ij}=[\partial V^{(3)}/\partial S_i \partial S_j]_{0,0,d}/V^0$, etc. First derivatives with respect to S_2 are zero.

$Y=1+C_3d^2+C_6d^3+C_{10}d^4$	(a)
$Y_3=2C_3d+3C_6d^2+4C_{10}d^3$	(b)
$Y_{22}=2C_3-6C_6d+4C_{10}d^2$	(c)
$Y_{33}=2C_3+6C_6d+12C_{10}d^2$	(d)
$Y_1=C_1+C_5d^2+C_9d^3-\gamma Y/2$	(e)
$Y_{13}=2C_5+3C_9d^2-\gamma Y_3/2$	(f)
$Y_{11}=2C_2+2C_8d^2-\gamma(Y_1+\gamma Y/2)$	(g)

calculate the energies of a series of C_{2v} structures, for which S_2 is zero but S_1 and S_3 are not zero, then we can determine all terms up to and including 5th order in S_i but we cannot determine 6th order because there are two terms, namely $(S_2^2+S_3^2)^3$ and $S_3^2(S_3^2-3S_2^2)^2$, which are of the same order, and therefore not distinguishable, for C_{2v} structures. To determine the coefficients of a polynomial beyond 5th order it would be necessary to have *ab initio* points for a lower symmetry at which both S_2 and S_3 are non-zero; these are at C_S points on the surface[2].

The ground state of O_3 dissociates to ground-state products according to the scheme

$$O_3(\tilde{X}, {}^1A_1) \rightarrow O_2(X, {}^3\Sigma_g^-) + O(^3P) \qquad 10.5$$

and the implication of this is that the ground-state surface is single-valued. There are, however, two indications that this may not be so. Calculations by Hay *et al.*[3] show that for a $D_{\infty h}$ structure with a bond length of 1.37 Å the ground-singlet state is ${}^1\Delta_g$ and this cannot correlate with ground-state products according to the Wigner–Witmer rules; to satisfy (10.5) the ground state would have to be ${}^1\Sigma_g^+$ and there is probably a crossing of ${}^1\Delta$ and ${}^1\Sigma^+$ states at some point or points of $C_{\infty v}$ symmetry on the surface. As ${}^1\Delta_g$ becomes ${}^1A_1$ plus ${}^1B_1$ in C_{2v} and ${}^1\Sigma_g^+$ is ${}^1A_1$, a crossing would give rise to a conical intersection of ${}^1A_1$ states in a linear configuration.

The second indication that the O_3 surface should not be treated as a single-valued function is that in the Hartree–Fock approximation, with electrons paired in orbitals, the D_{3h} and C_{2v} minima on the surface arise from different electron configurations, one being related to the other by a two-electron excitation. These two configurations are mixed in a CI scheme and it has been definitely established that both types of minima occur on the same surface, $^1A_1$ in C_{2v} or $^1A'$ in C_S. However, a scheme in which the O_3 surface is represented by the interaction of two diabatic surfaces has the merit of encompassing the intersection in linear configurations. The first potential functions produced for O_3 were single-valued[1,4,5], but a more elaborate two-valued surface has also been obtained[6]. The single-valued functions have the merit of being more economical when used in dynamical calculations[5,7].

By fitting the polynomial coefficients of the three-body term (10.4) to the C_{2v} spectroscopic data, as described earlier, a series of potentials can be obtained corresponding to different values of the range parameter γ (see Table 10.1). There are two features of the surface away from the minima which are important for dynamical studies on the O_3 surface. The first is the energy and geometry of the D_{3h} minimum and the second is the existence or not of a barrier on the reaction path for dissociation according to (10.5). Kinetic evidence is that there is little or no barrier to dissociation; an upper limit of 1 kcal mol^{-1} has been suggested[8], but it is more likely that no barrier exists.

A great deal of effort has been put into calculating an accurate energy for the D_{3h} minimum because of its possible role as an intermediate in the photochemical formation of O_3, and as an excited state of the molecule giving rise to the so-called Huggins band. The difference in energy between the D_{3h} and C_{2v} minima and the relative energies of both to the $O_2 + O$ dissociation limit is very sensitive to basis set and to the extent of configuration interaction in the wave functions. From some of the most extensive calculations, Burton[9] has concluded that the most likely value of the D_{3h} energy is 12 ± 2 kcal mol^{-1} above the C_{2v} minimum which is approximately 12 kcal mol^{-1} below the dissociation limit. The calculated bond length of the D_{3h} structure is 1.435 Å, which is larger than that in the C_{2v} molecule, 1.272 Å.

With a single-valued representation of the O_3 surface and a three-body polynomial to quartic terms it was not found possible to obtain a satisfactory energy for the D_{3h} minimum and to have a small barrier to dissociation. Low values of γ gave a D_{3h} minimum but a high dissociation barrier; high values of γ gave no barrier but no D_{3h} minimum. The simplest form of the three-body term which could be found to reproduce these important features of the surface was with a high γ and with the D_{3h} minimum imposed by subtracting a further three-body term whose maximum amplitude was centred at the position of the D_{3h} minimum and decayed to zero for distortions along the S_2 and S_3 coordinates[4]. The most recent variant of this function[5] is given in Table 10.2.

A two-valued representation of the O_3 surface has also been obtained[6] with one diabatic surface having the dissociation channel (10.5) and the other having

$$O_3(^1A_1) \rightarrow O_2(a, {}^1\Delta_g) + O(^1D) \qquad 10.6$$

The D_{3h} minimum is associated with the first of these (being $^1\Sigma_g^+$ in $D_{\infty h}$) and the C_{2v} minima with the second ($^1\Delta_g$ in $D_{\infty h}$). The diagonal terms in the V matrix had the same general form as the single-valued surface and the off-diagonal term was taken as

Table 10.2 A single-valued potential for O_3 (ref. 5). The S_i variables are defined in (10.1). Energies in eV, distances in Å.

$$V = \sum_{i=1,3} V^{(2)}(r_i) + V^{(3)}$$

$$V^{(2)}(r_i) = -5.213[1+3.7537r_i]\exp(-3.7537r_i), \quad r_i = R_i - 1.2074$$

$$V^{(3)} = [P(S_i)+G(S_i)][1-\tanh(4.4S_1/2)]$$

$$P(S_i) = 7.9920 + 6.1872S_1 + 12.4339S_1^2 - 15.1314(S_2^2+S_3^2) - 3.2418S_1(S_2^2+S_3^2) + 2.6323S_3(S_3^2-3S_2^2) + 12.4195(S_2^2+S_3^2)^2$$

$$G(S_i) = 3941.4(1+0.9919S_1)\left(1.2527 + \sum_{i=1,3}(R_i^2-R_j^2-R_k^2)/2R_jR_k\right)^5 \times \exp[-0.8(S_2^2+S_3^2)], \quad R_i^0 = 1.5698 \text{ Å}$$

$$V_{12} = \sin\alpha_1 \sin\alpha_2 \sin\alpha_3 V_{12}^{(3)} \qquad 10.7$$

where α_i are the internal angles of the triangle; this off-diagonal term preserves the permutation symmetry of the adiabatic surfaces and is zero for linear configurations.

In addition to the spectroscopic and *ab initio* data on the C_{2v} and D_{3h} minima there have been *ab initio* calculations on linear configurations of the molecule[3], and these can be used to determine the diabatic functions because in linear configurations these are non-interacting. The elements of the 2×2 V matrix were finally determined by optimizing calculated frequencies of the C_{2v} minima on the lower adiabatic surface to observed frequencies[6]. The potential function is defined in Table 10.3 and the calculated frequencies are compared with experiment in Table 10.4.

Figures 10.1 and 10.2 show contours of the ground-state potential and Figure 10.3 shows a schematic bending potential for C_{2v} structures. The diabatic and adiabatic surfaces differ in energy by approximately 0.5 eV at the angle of greatest interaction, which is 85°.

10.2 ClO_2 and HOF

The potential surface of ClO_2 is similar to that of SO_2 in having three minima, two of which are symmetry equivalent. There is, however, a difference in the data available to determine the potential function because both ClOO and OClO are species that have been investigated spectroscopically, and harmonic force fields have been determined for both. The symmetric species (C_{2v} symmetry) is kinetically stable at room temperature but Benson and Buss[10] have shown that the asymmetric form (C_S symmetry) is thermodynamically the more stable. However, this form has only been studied spectroscopically by the matrix isolation technique[11].

The ground state of the C_{2v} species is $^2B_1$, which reduces to $^2A''$ in C_S (ref. 12). The ESR spectrum of ClOO has been detected in an Ar matrix at 4°K and the

Table 10.3 Parameters of a two-valued O_3 potential[6]. The diagonal element V_{11} of the potential matrix corresponds to scheme (10.5) and V_{22} to scheme (10.6). The three-body polynomials have variables ρ_i and permutation symmetry is implied. The range function T, however, employs the totally symmetric variable S_1 defined by (10.1), and γ_1 is the exponent of this variable.

One-body term: $V^{(1)}_{22} = 1.958$ eV (Only one $O(^1D)$ energy appears in the dissociation channel (10.6) because $O_2(a, {}^1\Delta_g)$ dissociates to ground-state atoms)

Two-body terms:

Parameter	D_e/eV	R_e/Å	a_1/Å^{-1}	a_2/Å^{-2}	a_3/Å^{-3}
$V^{(2)}_{11}$	5.2136	1.2074	5.4447	7.7828	5.9920
$V^{(2)}_{22}$	4.2319	1.2157	3.9782	0.0	0.0

Three-body terms:

Parameter	$V^{(3)}_{11}$	$V^{(3)}_{22}$	$V^{(3)}_{12}$
C_i/Å^{-1}	0.2660	1.2350	−0.0162
C_{ii}/Å^{-2}	2.1987	−1.1245	−3.9548
C_{ij}	1.3910	3.8524	4.0472
C_{iii}/Å^{-3}	2.0814	−1.2469	0.0775
C_{iij}	−0.8893	0.0465	−0.0377
C_{ijk}	−0.0905	4.3230	0.2062
C_{iiii}/Å^{-4}	−0.3908	−0.0816	2.9954
C_{iiij}	1.1908	−2.3422	11.7870
C_{iijj}	−1.6388	2.6811	17.9732
C_{iijk}	−3.2080	1.2921	35.4934
γ_1/Å^{-1}	4.9443	6.6986	3.6183
R^0_i/Å	1.6941	1.8605	1.3458
V^0/eV	2.5680	3.6910	2.1656

Table 10.4 Comparison of observed vibration frequencies for O_3 (cm^{-1}) with those calculated from the potential given in Table 10.3. In reference 6 frequencies are also given for the ^{18}O isotopes of the molecule.

Level	Obs.	Calc.	Level	Obs.	Calc.
100	1103	1103	011	1726	1730
010	701	701	003	3046	3054
001	1042	1043	201	3186	3207
200	2201	2201	102	3084	3085
002	2058	2061	021	2410	2409
110	1795	1790	012	2726	2736
101	2111	2112	111	2785	2786

magnitude of the components of the Cl hyperfine coupling tensor suggest that it is a Π radical of symmetry $^2A''$ (ref. 13). *Ab initio* calculations at the SCF level support this assignment[14], although the convergence of the SCF wave function has been found to be difficult and other calculations suggest that the ground state

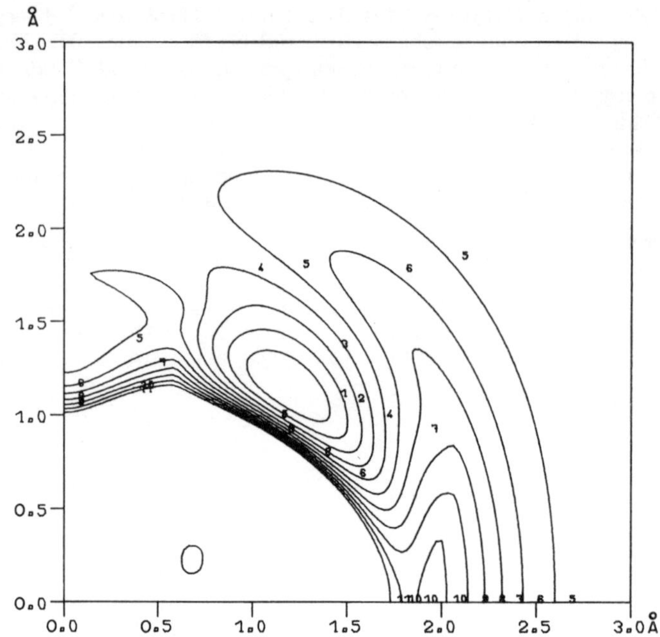

Figure 10.1 Contours of the O_3 potential defined in Table 10.3

is $^2A'$ (ref. 15). The difficult convergence suggests that both states are close in energy and, although we have produced a potential function on the assumption that both the ClOO and ClO_2 ground states are associated with the same potential surface, there is some doubt about this.

The energy, geometry and full harmonic force fields of ClOO and ClO_2 provide 17 pieces of data to determine the potential function, and we select the terms in the polynomial according to the criteria given in Appendix 2. If the three minima are located at points *a*, *b* and *c*, then the origin of the S_i coordinate system is the mid-point of the *a*–*b* line:

$$R_i^0 = \tfrac{1}{2}(R_{ia} + R_{ib}) \tag{10.8}$$

and the S_i coordinates are chosen such that S_1 is perpendicular to the plane containing *a*, *b* and *c*, S_2 is along the *a*–*b* line and *c* lies on the S_3 axis.

From the known bond lengths of ClO_2 ($R_{ClO} = 1.49$ Å, $\widehat{OClO} = 118.5°$) and ClOO ($R_{ClO} = 1.83$ Å, $R_{OO} = 1.23$ Å, $Cl\hat{O}O = 110°$) we deduce the following relationship between S_i and $\rho_i = R_i - R_i^0$

$$\begin{bmatrix} S_1 \\ S_2 \\ S_3 \end{bmatrix} = \begin{bmatrix} 0.5703 & 0.5913 & 0.5703 \\ 0.7071 & 0 & -0.7071 \\ -0.4181 & 0.8065 & -0.4181 \end{bmatrix} \begin{bmatrix} \rho_1 \\ \rho_2 \\ \rho_3 \end{bmatrix} \tag{10.9}$$

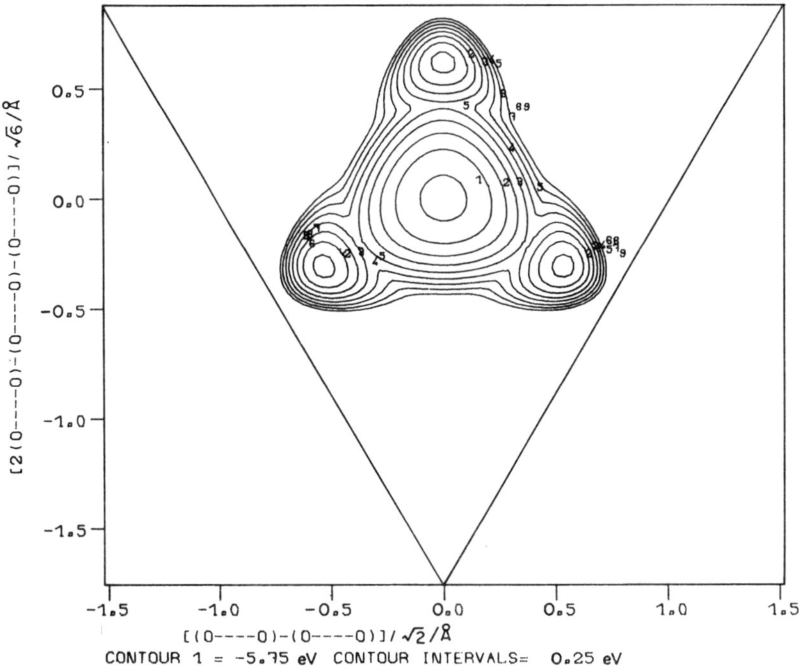

Figure 10.2 Contours of the O_3 potential defined in Table 10.3. The coordinates of a triangular graph are defined in Figure 7.8

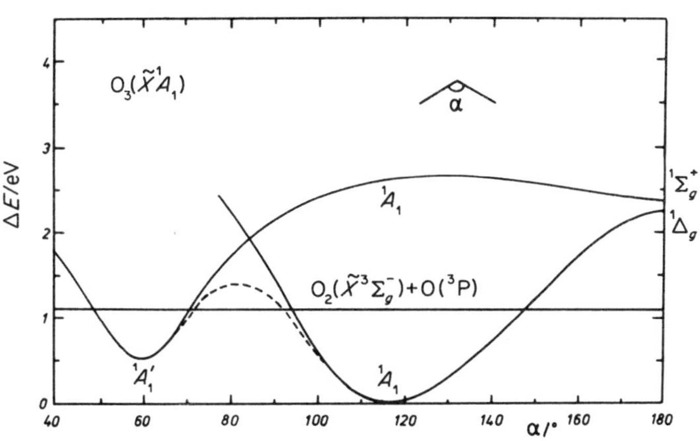

Figure 10.3 Schematic bending potential for O_3, maintaining C_{2v} symmetry. The diabatic potentials are shown as full curves and the adiabatic ground state as a dashed curve

$$\rho_i/\text{Å} = R_i(\text{Cl—O}) - 2.1800, \quad i = 1, 3$$

$$\rho_2/\text{Å} = R_2(\text{O—O}) - 1.2300$$

The three minima on the surface therefore have the following S_i coordinates:

$$a=(0,\lambda,0),\quad b=(0,-\lambda,0),\quad c=(0,0,\mu) \tag{10.10}$$

where

$$\lambda=\frac{1}{2}\left[\sum_{i=1,3}(R_{ia}-R_{ib})^2\right]^{1/2}=0.495\ \text{Å} \tag{10.11}$$

and

$$\mu=\frac{1}{2}\left[\sum_{i=2,3}(2R_{ic}-R_{ia}-R_{ib})^2\right]^{1/2}=1.650\ \text{Å} \tag{10.12}$$

The relationship between the terms in the three-body polynomial and the data associated with the two minima on the surface has been summarized in Table A2.3 in Appendix 2. We particularly stress the point that not all cubic and quartic terms can be determined, not S_1^3 for example; yet one 5th-power term, S_3^5, can be determined. In very general terms one can see that the position of the minima on the surface make for a greater functional complexity in the S_2, S_3 plane than in the S_1 direction.

The range function of the three-body term would be generally taken as a two-parameter function depending on both S_1 and S_3, the symmetric coordinates. However, as can be seen from (10.9), S_1 is almost totally symmetric in the three internal coordinates and is therefore a nearly exact measure of the perimeter of the three-atom triangle. Under these circumstances, one would expect to find that the range function is mainly dependent on S_1 and only weakly dependent on S_3, which is almost a pure shape variable. Indeed, a satisfactory potential function has been obtained on this basis[16].

A number of potential functions have been investigated for ClO_2 using differing amounts of data pertaining to the two minima. The most complete function[16], using all data, is given in Table 10.5, and representations of the potential are shown in Figures 10.4 and 10.5. The potential is found to have a barrier for the approach of Cl to O_2 which is at a maximum for collinear configurations, but there is no barrier for oxygen approaching ClO either to form ClOO or OClO.

An earlier version of the ClO_2 potential[17], but one not substantially different from that given in Table 10.5, was used to study the dynamics of the reaction[18]

$$\text{ClO}(^2\Pi)+\text{O}(^3\text{P})\rightarrow \text{O}_2(^3\Sigma_g)+\text{Cl}(^2\text{P}) \tag{10.13}$$

This is an exothermic reaction, and although there is a barrier in the exit channel, it is well below the energy of the reactants and does not significantly hinder the reaction. The calculated room temperature rate constant of 2.6×10^{11} cm^3 molecule^{-1} s^{-1} is within the variation of experimental measurements (1.0–5.2 of these units), although the most recent measurement is approximately twice that calculated from the surface[19]. As no information, either of a kinetic nature or from *ab initio* calculations, was used to determine the nature of the surface away from the minima, this result should be considered as very satisfactory. Such information, if used to obtain a better surface, would give greater confidence to the dynamical calculations.

An initial study[20] of the potential surface for HOF which was based only on the

Table 10.5 The many-body potential for ClO_2 (ref. 16). The three-body term is expressed in S_i variables.

Two-body terms:

Species	D_e/eV	R_e/Å	a_1/Å^{-1}
O_2	5.2130	1.2074	3.7537
ClO	2.8035	1.5660	3.2551

Three-body terms:

C_1/Å^{-1}	−0.1155	C_{2222}/Å^{-4}	2.4654
C_3	1.1817	C_{3333}	−2.9285
C_{11}/Å^{-2}	0.3824	C_{1333}	−2.4003
C_{22}	−0.5190	C_{2233}	2.0975
C_{33}	0.9375	C_{33333}/Å^{-5}	1.1703
C_{13}	−0.1931	V^0/eV	2.1536
C_{333}/Å^{-3}	0.4425	γ_1/Å^{-1}	3.0
C_{122}	0.0818	γ_2	0.0
C_{133}	4.6734	γ_3	0.0
C_{113}	1.1103	R_1^0(Cl—O)/Å	2.1800
C_{223}	−3.5377	R_2^0(O—O)	1.2300
		R_3^0(Cl—O)	2.1800

$S_1=0.5703\rho_1+0.5913\rho_2+0.5703\rho_3$
$S_2=0.7071\rho_1-0.7071\rho_3$
$S_3=-0.4181\rho_1+0.8065\rho_2-0.4181\rho_3$

spectroscopic properties of the equilibrium molecule suggested that there were two metastable minima on the surface corresponding to a bent species of formula HFO and a linear species of formula FHO. Although a species such as HFO contravenes the normal rules of valence, it should be remembered that valence is a rationalization of stable, not metastable, structures. Some stability for the species HFO is expected from the fact that in the radical OF the unpaired electron is distributed roughly 75% on the oxygen atom and 25% on the fluorine atom. A hydrogen atom approaching the fluorine should therefore experience some attractive force.

Calculations on both the metastable minima on the HOF surface were made using a double zeta basis, both with and without additional polarization functions, and various levels of CI[20]. These supported the existence of the metastable HFO predicting an energy of 4 kcal mol^{-1} below the $O(^1D)+HF(^1\Sigma^+)$ dissociation limit and geometrical parameters $R_{HF}=0.93$ Å, $R_{FO}=1.68$ Å, $H\hat{F}O=96°$. The linear structure was predicted to be a typical weak hydrogen-bonded species, just bound relative to $O(^1D)+HF(^1\Sigma^+)$, with $R_{HF}=0.92$ Å and $R_{OH}=2.09$ Å.

Like H_2O, the ground singlet state of HOF does not dissociate to ground-state products. The dissociation scheme which follows allows for either the O or the HF to be in an excited state:

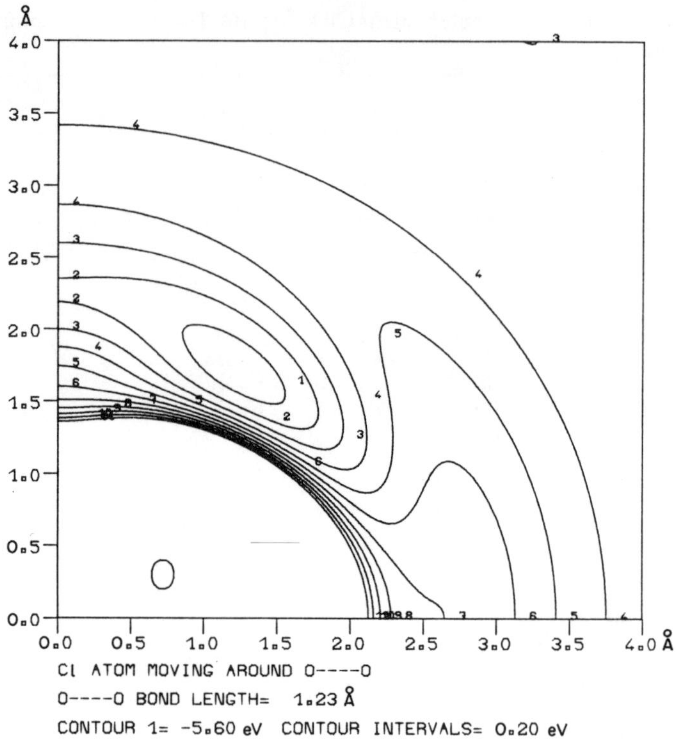

Figure 10.4 Contours of the ClO_2 potential of Table 10.5

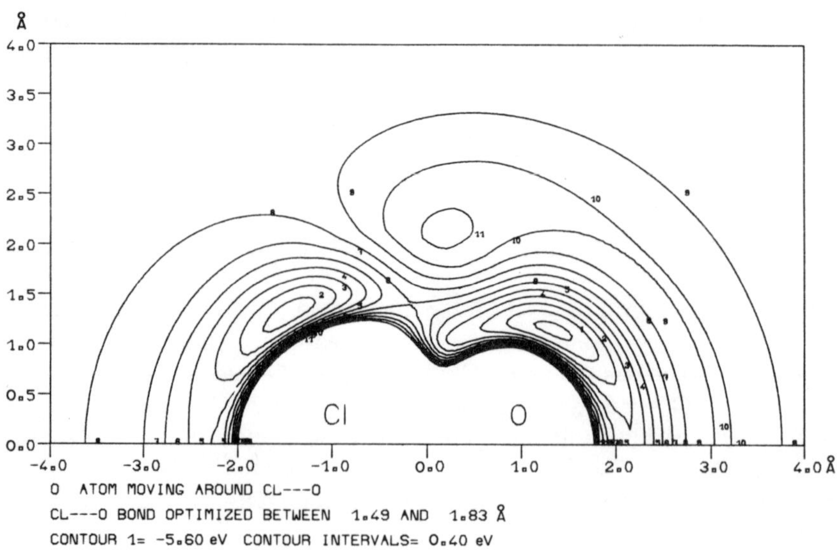

Figure 10.5 Contours of the ClO_2 potential of Table 10.5. The Cl—O bond length has been optimized at each position of the other atom so that the minima associated with the two isomers can both be obtained on the same figure

$$\mathrm{HOF}(^1A') \rightarrow \begin{cases} \mathrm{O}(^3\mathrm{P}) + \mathrm{HF}(^3\Sigma^-) & (i) \\ \mathrm{O}(^1\mathrm{D}) + \mathrm{HF}(^1\Sigma^+) & (ii) \\ \mathrm{H}(^2\mathrm{S}) + \mathrm{OF}(^2\Pi) & \\ \mathrm{F}(^2\mathrm{P}) + \mathrm{OH}(^2\Pi) & \end{cases} \tag{10.14}$$

and a single-valued surface has been determined which satisfies this, using the original Murrell–Sorbie procedure as described in Section 9.4. For a complete treatment of the surface one should derive a two-valued surface according to the method described in Chapter 9.

From the spectroscopically determined potential energy function for the ground state of HF and a calculated potential for the excited triplet state, it can be deduced that channel (*ii*) has the lower energy for $R_{HF} < 1.9148$ Å and each of the three minima on the surface satisfy this limit. Thus the three-body term for $R_{HF} < 1.9148$ Å can be determined in an analogous procedure to that used for ClO_2, and the three-body term in the region defined by $R_{HF} > 1.9148$ Å is determined by making the total potential and its derivatives continuous at the point at which the OH and OF bond lengths are the same as in equilibrium HOF.

A potential has been obtained[16] by fitting the energy, geometry and quadratic force field of HOF, and the energies and geometries of HFO and FHO deduced by *ab initio* calculations. With S_i coordinates chosen with respect to the three minima on the surface, as described in Section 4.3, it was found necessary to have three parameters γ in the range function to obtain a satisfactory surface. However, with a further translation of the origin to bring it closer to the HOF minimum, and a rotation about S_2, a single range parameter could be found which gave a satisfactory surface. The full potential determined in this way is given in reference 16. Examination of this potential shows that from the equilibrium structure HOF there is no barrier for removal of either H or F but there is a barrier to removal of H from HFO. The potential has been used to calculate the vibration frequencies of both HOF and the metastable hydrogen-bonded structure[16].

References

1. J. N. Murrell, K. S. Sorbie and A. J. C. Varandas, *Mol. Phys.*, **32**, 1359 (1976).
2. A. J. C. Varandas and J. N. Murrell, *Chem. Phys. Lett.*, **84**, 440 (1982).
3. P. J. Hay, T. H. Dunning and W. A. Goddard, *J. Chem. Phys.*, **62**, 391 (1975).
4. J. N. Murrell and S. C. Farantos, *Mol. Phys.*, **34**, 1185 (1977).
5. A. J. C. Varandas and J. N. Murrell, *Chem. Phys. Lett.*, **88**, 1 (1982).
6. S. Carter, I. M. Mills, J. N. Murrell and A. J. C. Varandas, *Mol. Phys.*, **45**, 1053 (1982).
7. A. J. Stace and J. N. Murrell, *J. Chem. Soc., Faraday Trans. 2*, **74**, 2182 (1978).
8. W. C. Gardiner, *Rates and Mechanisms of Chemical Reactions*, Benjamin, London, 1969.
9. P. G. Burton, *J. Chem. Phys.*, **71**, 961 (1979).
10. S. W. Benson and J. H. Buss, *J. Chem. Phys.*, **27**, 1382 (1957).
11. A. Arkell and I. Schwager, *J. Am. Chem. Soc.*, **89**, 5999 (1967).
12. D. Papousek and J. Pliva, *Collect. Czech. Chem. Commun.*, **29**, 1973 (1964).

13. F. J. Adrian, E. L. Cochran and V. A. Bowers, *J. Chem. Phys.*, **56**, 6251 (1972).
14. A. Hinchliffe, *J. Mol. Struct.*, **64**, 117 (1980).
15. J. L. Gole and E. F. Hayes, *Int. J. Quantum Chem.*, **3**, 519 (1970).
16. J. N. Murrell, S. Carter and I. M. Mills, *Mol. Phys.*, **37**, 1885 (1979).
17. S. Farantos, E. C. Leisegang, J. N. Murrell, K. Sorbie, J. J. C. Teixeira-Dias and A. J. C. Varandas, *Mol. Phys.*, **34**, 947 (1977).
18. S. C. Farantos and J. N. Murrell, *Int. J. Quantum Chem.*, **14**, 659 (1978).
19. M. S. Zahniser and F. Kaufman, *J. Chem. Phys.*, **66**, 3673 (1977).
20. J. N. Murrell, S. Carter, I. M. Mills and M. F. Guest, *Mol. Phys.*, **37**, 1199 (1979).

Chapter 11

Potential functions for ions

11.1 General principles

There are no fundamental differences between the potential energy surfaces of ions and those of neutral molecules, although there are important differences in the functional forms of the potential at large separations of the fragments. Neutral closed-shell fragments have a van der Waals interaction energy which is asymptotic to the function $-C_6R^{-6}$ but two ions interact with an energy C_1R^{-1}; an ion and a neutral (non-polar) closed shell fragment with $-C_4R^{-4}$, etc. However, in the short-range (valence) region of interaction there is no difference between neutral and ion interactions, although negative ions are generally less stable than neutral molecules or positive ions.

The reason why the potential functions of ions should be treated as a separate topic is that they are usually considerably more complicated than those of neutral systems, because the number of dissociation channels is approximately twice that of a similar neutral system as the charge can reside on either of two dissociation fragments.

We begin by considering a model triatomic ion ABC^+ for which the Wigner–Witmer rules allow dissociation to ground-state fragments. We therefore write the dissociation scheme as follows:

$$ABC^+ \rightarrow \begin{cases} \begin{cases} AB^+ + C \rightarrow \begin{cases} A^+ + B + C & (a) \\ A + B^+ + C & (b) \end{cases} \\ AB + C^+ \rightarrow \quad A + B + C^+ & (c) \end{cases} \\ \begin{cases} BC^+ + A \rightarrow \begin{cases} B^+ + C + A & (b) \\ B + C^+ + A & (c) \end{cases} \\ BC + A^+ \rightarrow \quad B + C + A^+ & (a) \end{cases} \\ \begin{cases} AC^+ + B \rightarrow \begin{cases} A + C^+ + B & (c) \\ A^+ + C + B & (a) \end{cases} \\ AC + B^+ \rightarrow \quad A + C + B^+ & (b) \end{cases} \end{cases} \qquad 11.1$$

The complication which this scheme suggests may or may not lead to a complication of many-valuedness in the potential energy function as will be indicated by the following examples.

If we are interested in the possibility of charge exchange according to the scheme

$$Ar^+ + Cl_2 \rightarrow Ar + Cl_2^+ \qquad 11.2$$

then we must examine the potential energy surfaces of $ArCl_2^+$. Reaction (11.2) is exothermic by the difference between the ionization potential of Ar (15.76 eV) and that of Cl_2 (11.50 eV), namely 4.26 eV. This means that the surfaces which are asymptotic to $Ar^+ + Cl_2$ are well above the surfaces which are asymptotic to $Ar + Cl_2^+$. The gap is a function not only of the Ar–Cl distances, but also of the Cl–Cl distance. As the ionization potential of Cl (13.01 eV) is greater than that of Cl_2, the gap is reduced to approximately 2.75 eV when the Cl–Cl distance is infinite. However, it is almost certain that the surfaces asymptotic to $Ar^+ + Cl_2$ are everywhere above those asymptotic to $Ar + Cl_2^+$ and hence reaction (11.2) requires a non-adiabatic process in order to proceed. The ground-state surface of $ArCl_2^+$ can therefore be represented by a single-valued function, although there are possible complications due to the degeneracy of the ground state of $Cl_2^+(^2\Pi_g)$.

For charge transfer between Xe^+ and Cl_2 the situation is different. The ionization potential of Xe is 3.6 eV lower than that of Ar, so that whilst it is still exothermic for Xe to capture an electron from Cl_2, it is endothermic for electron capture from Cl. There is therefore a crossing of the surfaces of Xe^+–Cl_2 and Xe–Cl_2^+ for infinite separation of the Xe and Cl_2, which becomes an avoided crossing at finite separations. This is illustrated in Figure 11.1. The representation of the ground-state surface of $XeCl_2^+$ must therefore be multi-valued[1].

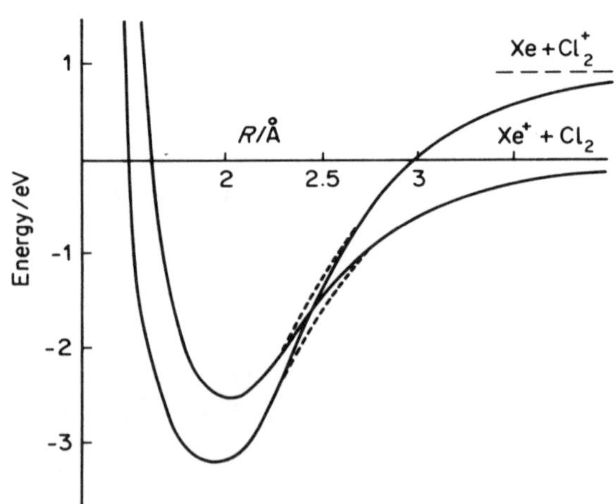

Figure 11.1 Diabatic potentials of $XeCl_2^+$ as a function of the Cl–Cl distance with Xe far removed. The dashed curves indicate the adiabatic potentials, their separation near the diabatic crosssng point being a function of the Xe–Cl_2 separation

Let us now look in more detail at the potential energy function appropriate for scheme (11.1). There must be at least three surfaces to satisfy this scheme because in the three-atom limit the positive charge can reside on any one of the three atoms. If any of these atoms or ions have degenerate states there will be more surfaces arising from the separated systems, but we are concerned now with a minimal description of the system. A particular state of a heteronuclear diatomic ion AB^+ will, on dissociation, leave the charge on either A or B. Thus if we wish to consider both the $A^+ + B$ and $A + B^+$ channels there must be two states of AB^+ to be considered in our scheme. The three separated atom limits in scheme (11.1) are labelled (*a*), (*b*) and (*c*), and we can write the many-body expression for the diagonal elements of the potential matrix as follows:

$$V_{aa} = V_{A^+}^{(1)} + V_{AB^+}^{(2)}(R_1) + V_{BC}^{(2)}(R_2) + V_{AC^+}^{(2)\prime}(R_3) + V_{aa}^{(3)}(R_1, R_2, R_3)$$

$$V_{bb} = V_{B^+}^{(1)} + V_{AB^+}^{(2)\prime}(R_1) + V_{BC^+}^{(2)}(R_2) + V_{AC}^{(2)}(R_3) + V_{bb}^{(3)}(R_1, R_2, R_3)$$

$$V_{cc} = V_{C^+}^{(1)} + V_{AB}^{(2)}(R_1) + V_{BC^+}^{(2)\prime}(R_2) + V_{AC^+}^{(2)}(R_3) + V_{cc}^{(3)}(R_1, R_2, R_3) \qquad 11.3$$

If the neutral atoms are in their ground states (energies taken as zero), then the one-body terms are the ionization potential of the atoms. One cannot say, in general, which of the positive ion diatomic states, one distinguished by a prime, has the lower energy. Sheet (*a*), for example, has the atom dissociation limit $A^+ + B + C$ and if A has a lower ionization potential than either B or C then the two-body terms appearing in V_{aa} will be ground-state potentials, that is, $V_{AB^+}^{(2)}$ will be a lower potential curve than $V_{AB^+}^{(2)\prime}$.

If the two-body terms are the exact adiabatic potentials of the diatomics then there can be no mixing of AB^+ in the absence of C. As there can be no mixing at infinite separations of A and B we conclude that the off-diagonal elements of the V matrix are three-body terms:

$$V_{ab} = V_{ab}^{(3)}(R_1, R_2, R_3), \quad \text{etc.} \qquad 11.4$$

As charge exchange is an overlap-dependent phenomenon we expect $V_{ab}^{(3)}$ to decay exponentially to zero as the atoms are separated; thus the simplest analytical form of the off-diagonal terms might be

$$V_{ab}^{(3)} = K \exp\left[-(\alpha_1 R_1 + \alpha_2 R_2 + \alpha_3 R_3)\right] \qquad 11.5$$

More complicated functions, as have been used in two-valued surfaces for neutral systems, may also be needed (see Chapter 9).

It is generally assumed that the many-body expansion will converge more slowly for ions or for neutral systems with polar fragments than for neutral, non-polar systems. This is because in the perturbation theory of intermolecular energies the most important non-pair-additive contributions are the induction energies, and these are large for charged or polar systems. Induction energy arises from the interaction of a static charge distribution on one fragment with the multipole moments that this induces on a neighbouring fragment. For example, a charge will induce a dipole on a fragment at distance R whose magnitude is proportional to R^{-2}. Charge-dipole energies are proportional to R^{-2}, so charge induced-dipole energies are proportional to R^{-4}, i.e. they are a quadratic function of the field at the fragment. If a fragment is surrounded by several charged species then the induced multipole moments will depend on the strength of the combined fields (vectorially combined) of all fragments, and the energy, being a quadratic function, is not pair additive. However, Claverie[2] has shown that the non-pair-additive induction energies contribute, in second-order perturbation theory, to the three-body energies but not to higher orders. Murrell *et al.*[3] have also shown that charge-transfer energies, which are also expected to be important for ionic or polar systems, contribute in second order of perturbation theory to three- and four-body energies but not to higher orders. Thus, although one expects three- and possibly four-body energies to be more important for ions and polar systems than for neutral non-polar systems, there is no reason to suppose that five-body and higher-order terms will be relatively more important. These conclusions have been confirmed by SCF calculations on the clusters $(Li_nCl)^{(n-1)+}$ by Bounds and Hinchliffe[4] and on fragments of the LiF crystal by Murrell *et al.*[3].

With the exception of H_3^+ which we discuss in the next section, there are few accurate potential functions for polyatomic ions. DIM surfaces have been obtained for NeH_2^+ (ref. 5), $NeHHe^+$ (ref. 6) and $NeHe_2^+$ (ref. 7) using an approach similar to that employed earlier for HeH_2^+ (ref. 8). Gittins and Hirst[9] reported that a potential function similar to that used by Sorbie and Murrell for H_2O could be fitted to *ab initio* calculations on the NH_2^+ surface and Herbst[10] has used a similar function for CH_2^+.

11.2 The H_3^+ surface

Although H_3^+ is the simplest commonly studied polyatomic molecule with respect to its electronic structure, its surface has, as we shall see, the complication of an avoided intersection. This is associated with the crossing of two diabatic surfaces which correspond to the structures H_2^+–H and H_2–H^+. Thus this system exemplifies our comment made at the beginning of this chapter that the potential surfaces of ions are generally more complicated than those of neutral systems.

H_3^+ and its isotopic variants are of considerable importance in astrophysics and their vibration–rotation spectroscopy has been studied both for transitions amongst the lowest vibrational states[11,12] and for transitions near the dissociation limit[13]. Not surprisingly, in view of its electronic simplicity, there have been many accurate studies of the H_3^+ potential surface and of the rotational–vibrational levels of the surface[14,15].

The first analytical potential for H_3^+ was by Giese and Gentry[16]. They used a sum of three diatomic H_2 potentials and varied the width, depth and position of the minimum according to the H_2–H^+ separation. Their parameters were fitted by a least-squares criterion to several sets of *ab initio* points with suitable adjustments to bring them into common accuracy.

Carney and Porter[17] at a similar time fitted CI calculations to a generalized

Morse function interpolation. Starting with the symmetry coordinates for an X_3 system (4.22), they transposed to spherical polar form (S, θ, ϕ), where

$$S_1 = S \cos \theta$$

$$S_2 = S \sin \theta \cos \phi$$

$$S_3 = S \sin \theta \sin \phi \qquad 11.6$$

and by examining cuts of the surface for fixed values of θ and ϕ they parameterized a generalized Morse function in S for each of these cuts and then interpolated these parameters in the (θ, ϕ) space.

In a later and more accurate study of the H_3^+ surface, Schinke *et al.*[18] calculated 650 points on the surface by a complete CI calculation from a (5s, 3p) Gaussian basis (42 basis functions, 606 A' configurations). These points, together with some results from the perturbation theory of the H^+–H_2 interaction at large separations, were fitted by analytical forms in five different regions of configuration space. The functions and variables used were various: in some regions scattering coordinates and the expansion (4.2) was used, in others the variables were the internuclear distances and the functions were products of polynomials up to sixth order and exponentials.

The surface intersection was tackled by the method used earlier by Giese and Gentry[16] which was to fit the lower sheet of the adiabatic surfaces for small separations of H^+ and H_2 (R in the scattering coordinates being less or equal to $4a_0$) but to fit the diabatic surface for $R > 4a_0$. The justification is that for $R > 4a_0$ the adiabatic splitting of the two surfaces is very small (almost cusp-like behaviour) and as the potential functions were being developed for rotational inelastic scattering and not for charge-exchange or reactive collisions, this near crossing should be treated as a true crossing.

Tennyson and Sutcliffe[15] used the potential of Schinke *et al.* for a study of the rotational–vibrational levels of H_3^+ in scattering coordinates and it was therefore necessary to expand the potential in Legendre functions (4.2), thirty terms being needed to provide the required accuracy.

The complications found necessary to Schinke *et al.* to provide a good fit to the partly adiabatic, partly diabatic lower sheet of the H_3^+ two-valued surface, suggests that it is in fact simpler to use a many-valued representation in which the elements of the potential matrix are determined by the procedure described in Chapter 9. There is the additional advantage that the surface can then be used for reaction and charge exchange as well as for rotational scattering and spectroscopy. It is almost certainly necessary to use such a function to describe the rotational–vibrational states near the dissociation limit[13].

A DIM treatment of the H_3^+ surface was obtained in early work by Ellison *et al.*[19]. This was based on the three valence bond structures that can be formed by coupling two hydrogen 1s orbitals in a Heitler–London electron pair bond. The DIM matrix elements reduce to combinations of the energies of the $H_2(X, {}^1\Sigma_g^+)$, $H_2^+(X, {}^2\Sigma_g^+)$ and $H_2^+(A, {}^2\Sigma_u^+)$ electronic states and overlap integrals. Preston and Tully[20] used the zero-overlap version of DIM and carried out trajectory calculations on the lowest two eigenvalues. However, it was later shown by Schinke *et al.*[18] that the second of these two eigenvalues is in poor agreement with *ab initio* calculations for close approaches of all three atoms. Top and Baer[21,22] also used the DIM surface for a study of elastic, reactive and charge-exchange processes.

If we return now to the general three-valued representation of the ABC^+ potentials, (11.3) and (11.4), then for H_3^+ the two-body terms will be the potentials of the ground states of H_2 and H_2^+ and the excited state of H_2^+; the same terms that enter the DIM matrix. It can be seen, moreover, that on removing any one atom from the other two, the three diagonal elements reduce individually to one of these three potentials. We are not, however, generally interested in a surface which dissociates to the repulsive state of H_2^+ as it lies above the $2H + H^+$ limit. It follows that one sheet of the three-valued surface will, for most configurations, be well above the other two and this will not be important for low-energy dynamics.

The reduction of the three-valued surface to one single-valued and one two-valued surfaces cannot, however, be made exactly in any simple form. For D_{3h} configurations of H_3^+ the matrix representation of the surface, given by equations (11.3) and (11.4), takes the form

$$V = \begin{bmatrix} A & \alpha & \alpha \\ \alpha & A & \alpha \\ \alpha & \alpha & A \end{bmatrix} \qquad 11.7$$

which has eigenvalues $A + 2\alpha$, $A - \alpha$, $A - \alpha$.

Simple MO theory shows that the ground state of H_3^+ in D_{3h} configurations is non-degenerate (its energy will therefore correspond to $A + 2\alpha$) and the first excited singlet state is doubly degenerate ($A - \alpha$). Moreover, the excited state will exhibit a Jahn–Teller distortion and its potential must be represented by a two-valued surface with a conical intersection (as is the ground state of H_3). On distortion from D_{3h} configurations, and for large separations of one atom from the other two, we know that the ground state is crossed diabatically by one of the excited states. Thus an exact representation of the surfaces for all configurations cannot be reduced from a three-valued problem. Either there is an intersection in the excited states or there is an intersection of the ground and one excited state.

Carter[23] has derived an approximate two-valued representation of the H_3^+ surface by taking two diabatic states, one of which dissociates always to $H_2(X, {}^1\Sigma_g^+) + H^+$ and the other always to $H_2^+(X, {}^2\Sigma_g^+) + H$. From the analysis of the problem just given, we expect this basis to be adequate for the ground and first excited states except for the excited state at or near D_{3h} configurations. The diagonal elements of the potential matrix will be

$$\left.\begin{aligned} V_{aa} &= V_{\mathrm{HH}}^{(2)}(R_1) + V_{\mathrm{HH}}^{(2)}(R_2) + V_{\mathrm{HH}}^{(2)}(R_3) + V_{aa}^{(3)} \\ V_{bb} &= V_{\mathrm{HH^+}}^{(2)}(R_1) + V_{\mathrm{HH^+}}^{(2)}(R_2) + V_{\mathrm{HH^+}}^{(2)}(R_3) + V_{bb}^{(3)} \end{aligned}\right\} \qquad 11.8$$

where the zero of energy is taken as the $2H + H^+$ dissociation limit. The off-diagonal term will be a three-body term $V_{ab}^{(3)}$.

The parameters of the potential have been determined[22] by a least-squares fitting to *ab initio* data on the ground and first excited state obtained from Schinke *et al.*[18]. Initial fitting to the diagonal (diabatic) potentials was made using the data from points which were well removed from the diabatic intersection. Points on the excited state surface in excess of 10 eV above the asymptote $2H+H^+$ were ignored (in accordance with the acknowledged failure of the scheme in the high-energy region). The final fitting to obtain refined diagonal elements and the off-diagonal element was made with the inclusion of points near the diabatic crossing. Table 11.1 gives the parameters of this two-valued potential.

Table 11.1 Two-valued representation of the H_3^+ ground-state surface[23]. For reasons given in the text the second eigenvalue of this matrix will not be a good representation of the excited-state surface in all configurations. The three-body term is written in ρ_i variables with permutation symmetry assumed.

Two-body terms:

Species	D_e/eV	R_e/Å	a_1/Å^{-1}	a_2/Å^{-2}	a_3/Å^{-3}
$H_2(^1\Sigma_g^+)$	4.7501	0.7417	3.8795	3.7409	3.2546
$H_2^+(^2\Sigma_g^+)$	2.7995	1.0480	2.5097	1.2706	1.1913

Three-body terms:

Parameter	$V_{aa}^{(3)}$	$V_{bb}^{(3)}$	$V_{ab}^{(3)}$
V^0/eV	3.6683	10.2785	0.3878
R^0/Å	0.9249	1.5358	1.1783
γ_1/Å^{-1}	0.8907	1.5027	2.1492
C_1/Å^{-1}	−0.1187	0.0393	0.3192
C_{11}/Å^{-2}	0.0905	−0.0035	−0.7084
C_{12}	−0.2662	0.0711	−0.2253
C_{111}/Å^{-3}	0.0737	0.0269	0.9527
C_{112}	−0.0432	0.0078	0.1778
C_{123}	0.2129	0.1452	0.0597
C_{1111}/Å^{-4}	−0.0482	0.0097	−0.5182
C_{1112}	0.0795	−0.0318	−0.1000
C_{1122}	−0.0663	0.0234	0.1582
C_{1123}	−0.1052	0.0305	0.0426

Figure 11.2 shows contours of the diabatic potential V_{aa}, defined by equation (11.8). This potential is associated with dissociation to H^+ plus H_2. Figure 11.3 shows the lowest adiabatic sheet of the two-valued surface, and a comparison with Figure 11.2 shows that the ground state is predominantly V_{aa} in the region of the equilibrium distance, the model adopted by other workers. The intersection with the second adiabatic sheet occurs in D_{3h} configurations with $R_{HH}=2.3$ Å and, as shown in Figure 11.4, the surface is complicated in this region.

Contours of the second adiabatic potential that is obtained from Table 11.1 are

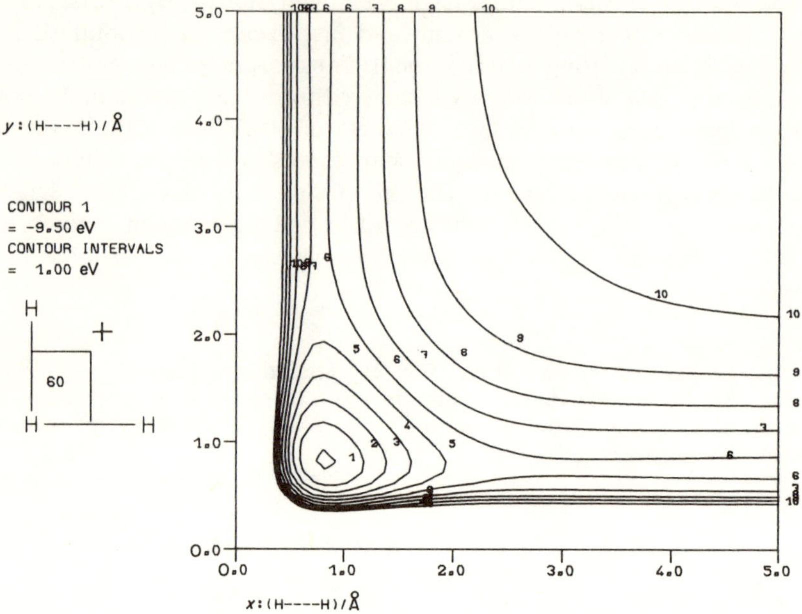

Figure 11.2 Contours of the diabatic potential V_{aa} of H_3^+ defined by equation (11.8) and Table 11.1, for a bond angle of 60°

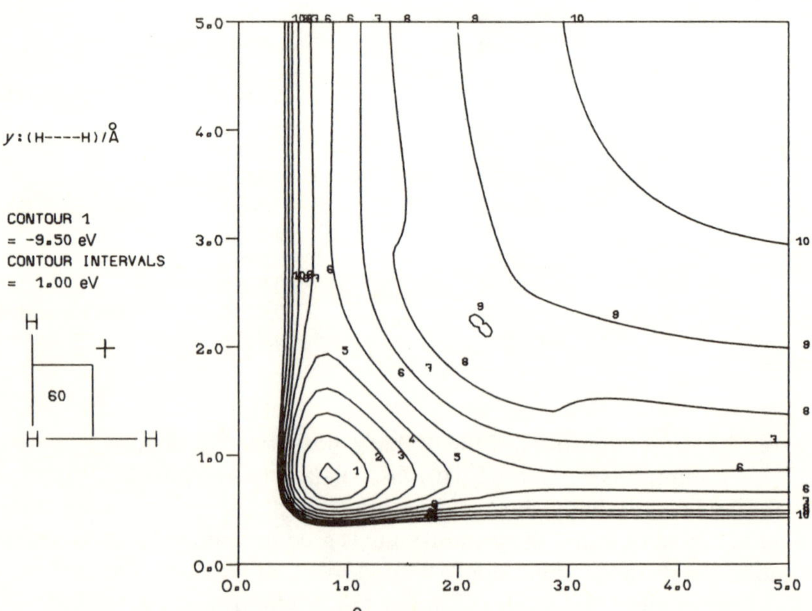

Figure 11.3 Contours of the lower adiabatic sheet of the H_3^+ potential of Table 11.1, for a bond angle of 60°

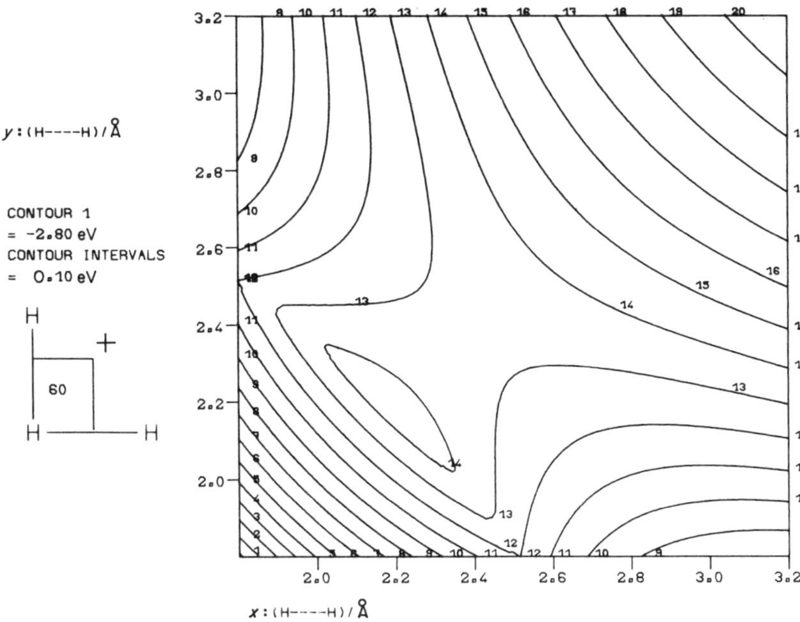

Figure 11.4 Expanded version of Figure 11.3 in the region of the crossing point, which is at H–H = 2.3 Å

not shown because, as explained in the text, the H_3^+ surface should be three-valued, and the second sheet in our model is not reliable in all space. Figure 11.5 shows another view of the lower adiabatic potential, and for most of the region shown this can be considered as the potential of H^+ moving around H_2.

Note added in proof. A Sorbie–Murrell potential for HeH_2^+ has recently been obtained[24] by fitting very accurate *ab initio* data of McLaughlin and Thompson[25]. Classical trajectories on this surface[26] give excellent agreement with experimental data for the reaction

$$He + H_2^+ \rightarrow HeH^+ + H$$

References

1. J. N. Murrell, *Chem. Phys. Lett.*, **93**, 521 (1982).
2. P. Claverie, in *Intermolecular Interactions from Diatomics to Biopolymers*, Ed. B. Pullman, Wiley–Interscience, New York, 1978.
3. J. N. Murrell, J. Tennyson and M. A. Kamel, *Mol. Phys.*, **42**, 747 (1981).
4. D. G. Bounds and A. Hinchliffe, *Mol. Phys.*, **40**, 989 (1980).
5. E. F. Hayes, A. K. Q. Sin, F. M. Chapman and R. L. Matcha, *J. Chem. Phys.*, **65**, 1901 (1976).
6. R. L. Matcha, B. M. Pettitt, P. F. Meier and P. Pendergast, *J. Chem. Phys.*, **69**, 2264 (1978).
7. J. Kendrick and P. J. Kuntz, *J. Chem. Phys.*, **70**, 736 (1979).
8. P. J. Kuntz, *Chem. Phys. Lett.*, **16**, 581 (1972).

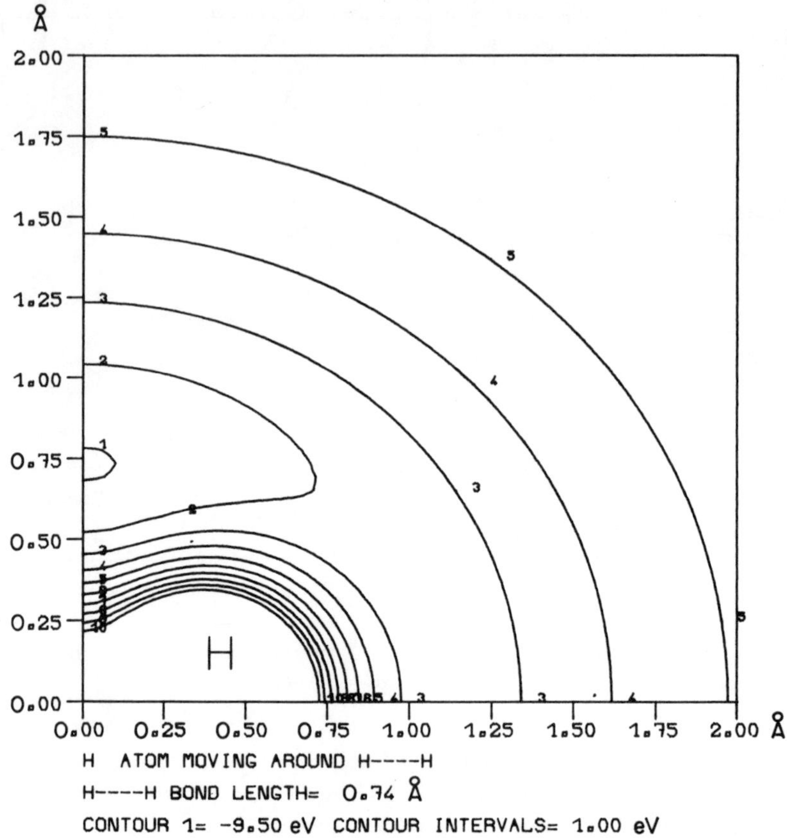

Figure 11.5 Contours of the lower adiabatic sheet of the H_3^+ potential for one atom moving around the other two at a diatomic bond length equal to that in the ground state of H_2

9. M. A. Gittins and D. M. Hirst, *Faraday Discuss. Chem. Soc.*, **62**, 67 (1977).
10. E. Herbst, *Chem. Phys. Lett.*, **47**, 517 (1977).
11. T. Oka, *Phys. Rev. Lett.*, **45**, 531 (1980).
12. J. T. Shy, J. W. Farley, J. W. Lamb and W. H. Wing, *Phys. Rev. Lett.*, **45**, 535 (1980).
13. A. Carrington, J. Buttenshaw and R. Kennedy, *Mol. Phys.*, **45**, 753 (1982).
14. G. D. Carney, *Chem. Phys.*, **54**, 103 (1980), and references cited therein.
15. J. Tennyson and B. T. Sutcliffe, *Mol. Phys.*, **51**, 887 (1984).
16. G. F. Giese and W. R. Gentry, *Phys. Rev. A*, **10**, 2156 (1974).
17. G. D. Carney and R. N. Porter, *J. Chem. Phys.*, **60**, 4251 (1974).
18. R. Schinke, M. Dupuis and W. A. Lester, *J. Chem. Phys.*, **72**, 3909 (1980).
19. F. O. Ellison, N. T. Huff and J. C. Patel, *J. Am. Chem. Soc.*, **85**, 3544 (1963).
20. R. K. Preston and J. C. Tully, *J. Chem. Phys.*, **54**, 4297 (1971); **55**, 562 (1971).
21. Z. H. Top and M. Baer, *Chem. Phys.*, **25**, 1 (1977).
22. M. Baer, *Top. Curr. Phys.*, **33**, 117 (1983).
23. S. Carter, unpublished results.
24. T. Joseph, PhD thesis, Indian Institute of Technology, Kanpur, India, 1983.
25. D. R. McLaughlin and D. L. Thompson, *J. Chem. Phys.*, **70**, 2748 (1979).
26. T. Joseph and N. Sathyamurthy, *J. Chem. Phys.*, **80**, 5332 (1984).

Chapter 12

Tetra-atomic molecules

12.1 General strategy

The increase in complexity of a potential energy surface on going from tri- to tetra-atomic systems makes it a formidable problem to find a potential function which is accurate over the whole configuration space. The two features which present particular difficulty are the increase in the number of stationary points on the surface which are associated with metastable structures and the increase in the number of intersections with other surfaces.

Some of the popular approaches to triatomic surfaces such as spline fitting have not been extended to tetra-atomics and perhaps cannot be extended. LEPS functions have been derived and used for the dynamics of a few systems[1,2] but they show up badly when tested by *ab initio* or empirical data. The DIM method is in principle extendable to tetra-atomic systems[3] but there have been few applications[4]. The problem here arises from the large number of functions that are needed to provide a good basis for the whole surface. This is more difficult than in the normal valence bond theory of chemical bonding because for that one is usually only interested in a basis which gives a good representation of the normal chemical structure.

A similar difficulty arises in any other method of splitting up the Hamiltonian, i.e. in a generalized molecules-in-molecules approach, if the Hamiltonian matrix is to be based on wave functions. A molecules-in-molecules treatment of H_4 has given poor results for high symmetry D_{3h} configurations[5].

In contrast, the many-body expansion method is extendable to tetra-atomic systems and the effort to do this is not significantly greater than that needed to obtain the potential functions of the four triatomic components of the system. Moreover, these triatomic potentials can be used to construct the potential of more than one tetra-atomic system. However, we are not wishing to over-state our case. The problem of surface intersections that we have already mentioned is unlikely to be solved exactly for most tetra-atomic systems, so that in practice the many-body expansion method will usually be used to represent only the most important regions of the surface.

Generally there are insufficient data on which to base an $n \times n$ matrix representation of an n-valued surface if n is greater than 2 and even if this representation could be constructed it might be difficult to use in a dynamical calculation. In Section 12.4 we show, with the example of H_2O_2, how the method developed in Section 9.4 for approximate single-valued representations of multi-valued surfaces can be extended to tetra-atomic systems.

The six internuclear distances R_i are an independent set of coordinates which define the geometry of a tetra-atomic system except for its chirality. As the potential energies of chiraly related configurations are identical the coordinates R_i are sufficient to define fully the potential energy function.

For triatomic systems the construction of the three-body term is greatly simplified by using S_i coordinates which are an orthogonal transformation of the displacements $\rho_i = R_i - R_i^0$ from a reference structure (R_i^0). The S_i coordinates are defined by the permutation symmetry of the system or, in the absence of symmetry, they can be selected so that minima on the potential surface lie on the S_i axes. The transformation matrices are defined in Chapter 4 and Appendix 2.

For tetra-atomic systems permutation symmetry can again be used. The reference configuration must be one for which all permutationally equivalent R_i are equal, for then the potential function will be expressed in totally symmetric combinations of S_i coordinates. For an A_4 system the reference is a tetrahedron and the S_i symmetry coordinates are given by (4.31). For an A_3B system the reference configuration is C_{3v} or D_{3h} and the A–A and A–B distances are transformed into separate irreducible representations of S_3 [equation (4.22)]. For an A_2B_2 system the reference is C_{2v} or D_{2h} and the four A–B distances are transformed as in (4.29). Finally, for an A_2BC system the reference configuration must be C_{2v} or C_s and the AB and AC distances form two separate S_i sets as in equation (4.34).

There may be some advantages in taking combinations of ρ_i above those imposed by symmetry, for example so that stationary points on the surface lie on S_i axes. However, unlike the case of a triatomic surface (Appendix 2), there are no general formulae we wish to propose for this task.

For triatomic molecules spectroscopists almost always use a coordinate system of two bond lengths and the included angle, and the transformation between this and the R_i coordinates has been treated generally in Appendix 1. For tetra-atomic molecules the favoured spectroscopic coordinates depend on the geometry of the stable configuration. For example, for NH_3 the coordinates are usually (but not uniquely) the three N–H bond lengths and the three angles $H\hat{N}H$. For C_2H_2 they are usually the three bond lengths, two $H\hat{C}C$ angles and the dihedral angle between the two HCC planes. Transformation between the spectroscopic variables and the R_i variables requires complicated algebra.

12.2 The ground-state surface of H_2CO

Formaldehyde is probably the most intensively studied tetra-atomic molecule of the A_2BC class in respect of its ground-state surface. The main feature of this surface is a deep minimum corresponding to a planar equilibrium structure with C_{2v} symmetry, the dimensions of which are given in Table 12.1 and Figure 12.1.

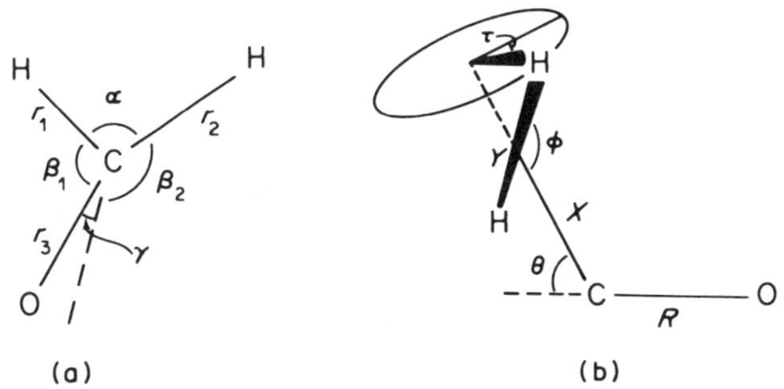

Figure 12.1 Definition of variables used to define the geometries of the equilibrium structure and the $H_2 \cdots CO$ saddle point

There are two further features of the formaldehyde surface that are important. The first is a metastable trans-planar hydroxycarbene structure (HCOH), which *ab initio* calculations[6] predict to be 2.22 eV higher in energy than that of the equilibrium configuration. The second is a substantial barrier (saddle point) to the formation of H_2CO from H_2 and CO. This process is exothermic by 0.31 eV, but there is no evidence of the reaction proceeding, even in the presence of a catalyst. *Ab initio* calculations[7] suggest that the saddle point, whose structure is given in Table 12.1 and Figure 12.1, has an energy 4.3 eV above the equilibrium configuration.

Adiabatic dissociation of the ground state of formaldehyde which conserves electron spin leads to the following dissociation products:

$$
H_2CO(\tilde{X}, {}^1A_1) \rightarrow
\begin{array}{ll}
H(^2S) + HCO(\tilde{X}, {}^2A') & \\
O(^3P) + CH_2(\tilde{X}, {}^3B_1) & \\
\left\{\begin{array}{ll} C(^3P) + H_2O(\tilde{a}, {}^3B_1) & (i) \\ C(^1D) + H_2O(\tilde{X}, {}^1A_1) & (ii) \end{array}\right. & \\
H_2(X, {}^1\Sigma_g^+) + CO(X, {}^1\Sigma^+) & \\
CH(X, {}^2\Pi) + OH(X, {}^2\Pi) &
\end{array}
\qquad 12.1
$$

The only channel which does not produce both fragments in their ground states is $C + H_2O$. Although the $H_2O(\tilde{a}, {}^3B_1)$ state is repulsive it is only when both OH bonds are short that the energy gap between this state and the singlet ground state is large. Thus for most of the surface which is relevant to the spectroscopy and dynamics of H_2CO it is the triplet–triplet combination labelled (*i*) in scheme (12.1) that has the lower energy. There are probably also some configurations of CH_2 for which the lowest singlet state $CH_2(\tilde{a}, {}^1A_1)$ is lower in energy than the triplet ground-state $CH_2(\tilde{X}, {}^3B_1)$ but the energy gap is unlikely to be so large that it compensates for the additional energy needed to promote the oxygen atom to $O(^1D)$, hence the singlet–singlet pairing in this channel is unlikely to be relevant.

A potential surface has been derived for H_2CO based on scheme (12.1) but with only channel (*i*) allowed at the $C + H_2O$ limit. This surface should be accurate for most configurations of interest to the dynamics of formaldehyde but it would not be suitable for any process which samples a region in which both OH bonds are short.

Potentials for the triatomic fragments can be obtained by the techniques described in Chapters 6 and 7. These provide the one-, two- and three-body terms needed for the H_2CO potential and these are given in Table 12.3. Before discussing the method by which the four-body term is obtained it is instructive to examine the many-body potential for the molecule without the four-body term. Some properties of this truncated potential have been calculated[8] and are compared with experimental and *ab initio* quantities in Table 12.1. Apart from an underestimation of the atomization energy of formaldehyde by 3.4 eV, the results are fairly encouraging. In particular, the equilibrium configuration is predicted to be planar with reasonably acceptable dimensions.

Table 12.1 Geometries and energies of the equilibrium configuration and $H_2\cdots CO$ saddle point predicted by the two- plus three-body potential. ΔE is the energy relative to the equilibrium structure.

(*a*) Geometry and energy of the equilibrium configuration (see Figure 12.1a)

	Exptl.[9,10]	2- + 3-body[8]
$r_1 = r_2$/Å	1.12	1.16
r_3/Å	1.21	1.27
α/°	116.0	124.1
$\beta_1 = \beta_2$/°	122.0	117.9
D_e/eV	16.28	12.88

(*b*) Geometry and energy of the $H_2\cdots CO$ saddle point (see Figure 12.1b)

	Ab initio[7]	2- + 3-body[8]
R/Å	1.22	1.20
X/Å	1.28	1.82
Y/Å	1.32	1.87
θ/°	50.0	20.0
ϕ/°	122.0	130.0
τ/°	0.0	90.0
ΔE/eV	4.3	1.38

Table 12.2 shows the force field of the equilibrium structure that is predicted from the two-body plus three-body terms. There is a 30% underestimation of the C–O stretching force constant and a 25% overestimation of the symmetric C–H stretching force constant, but the general balance is good. It should also be stressed that this truncated potential reproduces the energies of all the dissociation fragments correctly. These energies are summarized in Figure 12.2.

Figure 12.3 shows contours of the truncated potential for a hydrogen atom moving around a fixed HCO fragment, all four atoms being in one plane. The most noteworthy point is that there are secondary minima corresponding to HCOH metastable structures. The optimum HCOH structure is only slightly

Table 12.2 Force-field variables for the H_2CO equilibrium structure, and bond displacement variables for the four-body term in the many-body expansion, $\rho_i = R_i - R_i^0$. Variables are labelled by their symmetries in C_{2v}.

Force-field variables (Figure 12.1a)	Bond-displacement variables
$Q_1(A_1) = (\Delta r_1 + \Delta r_2)/\sqrt{2}$	$S_1(A_1) = \rho_1 + \rho_2$
$Q_2(A_1) = \Delta r_3$	$S_2(A_1) = \rho_3$
$Q_3(A_1) = (2\Delta\alpha - \Delta\beta_1 - \Delta\beta_2)/\sqrt{6}$	$S_3(A_1) = \rho_6$
$Q_4(B_1) = \gamma$	$S_4(A_1) = \rho_4 + \rho_5$
$Q_5(B_2) = (\Delta r_1 - \Delta r_2)/\sqrt{2}$	$S_5(B_2) = \rho_1 - \rho_2$
$Q_6(B_2) = (\Delta\beta_1 - \Delta\beta_2)/\sqrt{2}$	$S_6(B_2) = \rho_4 - \rho_5$

Harmonic force field in Q_i variables

	Exptl.[11]	2-body + 3-body[8]
F_{11}/aJ Å^{-2}	4.963	6.555
F_{22}	12.903	8.975
F_{55}	4.852	4.432
F_{12}	0.739	1.111
F_{13}/aJ Å^{-1}	0.077	1.445
F_{23}	−0.408	−0.202
F_{56}	0.171	−0.401
F_{33}/aJ	0.570	1.586
F_{44}	0.403	0.207
F_{66}	0.833	1.142

$R_1 = R_{CH}$, $R_2 = R_{CH'}$, $R_3 = R_{CO}$, $R_4 = R_{OH}$, $R_5 = R_{OH'}$, $R_6 = R_{HH}$

higher in energy than the most stable configuration, but this fault can be attributed to the poor energy of the latter. The predicted HCOH geometry is fairly good with respect to bond length but there is a 10° dihedral rotation from the *cis* configuration whereas the best *ab initio* calculation predicts a *trans* structure.

The truncated potential also predicts a reasonable geometry for the $H_2\cdots CO$ saddle point (Table 12.1). The energy of this, relative to the equilibrium structure, is much too low but that is again a result of the poor energy of the latter. To summarize: the truncated many-body potential for H_2CO, which neglects the four-body term, appears to be qualitatively satisfactory over large regions of the surface so that whilst it cannot be said that the four-body term is unimportant, it may be regarded as the fine-tuning term of the potential.

In reference 8, a four-body term for H_2CO was obtained by the methods described in Chapter 7 to reproduce exactly the spectroscopic parameters of the equilibrium structure, and this term also improved the surface in the HCOH and $H_2\cdots CO$ regions. However, the information available from *ab initio* calculations on these regions was not used in the optimization of the function. For the kinetic and photochemical processes involving the ground-state surface these regions are very important, particularly the height of the $H_2\cdots CO$ saddle point[7]. An improved surface which encompassed these data in the optimization routine has

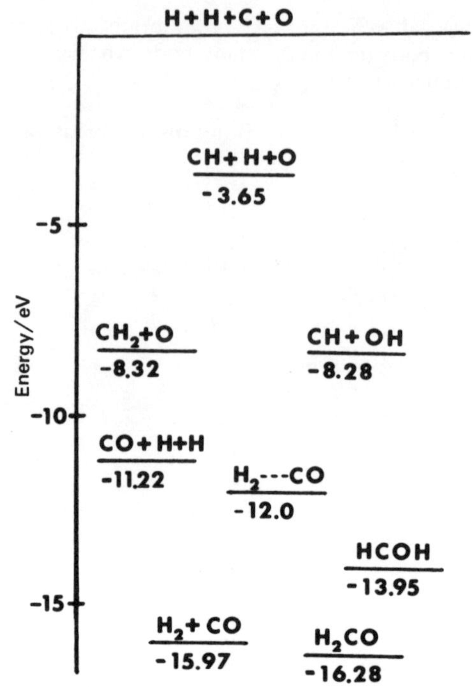

Figure 12.2 Relative energies of equilibrium H_2CO and its dissociation products which are reproduced exactly by the potential of Table 12.3. The relative energies of the *trans*-HCOH and the $H_2\cdots CO$ transition state predicted by the potential of reference 8 are also shown

been derived[12] and this additionally made use of better three-body terms than in the earlier work.

The reference structure for the four-body term was taken as the experimental equilibrium structure of H_2CO. The symmetry coordinates for displacement from this structure are defined in Table 12.2. S_1, S_2, S_3 and S_4 are all symmetric to permutation of the hydrogen atoms, S_5 and S_6 are antisymmetric and a suitable four-body term can therefore be expressed as

$$V^{(4)} = \prod_{i=1}^{4} [1 - \tanh(\gamma_i S_i/2)] P(S_j) \qquad 12.2$$

where the polynomial $P(S_j)$ takes any power of S_1, S_2, S_3 and S_4, but only even powers of S_5 and S_6. The complete quadratic polynomial contains a total of 18 independent terms.

We now consider the data which we wish to reproduce in two stages. The first set of data is that pertaining to the equilibrium configuration, namely its energy, geometry and harmonic force field as given in Tables 12.1 and 12.2. These make 14 pieces of data in all. Comparing the Q_i (spectroscopic) coordinates in Table 12.2 with the S_j (polynomial) coordinates, also shown, we see that only three Q_i

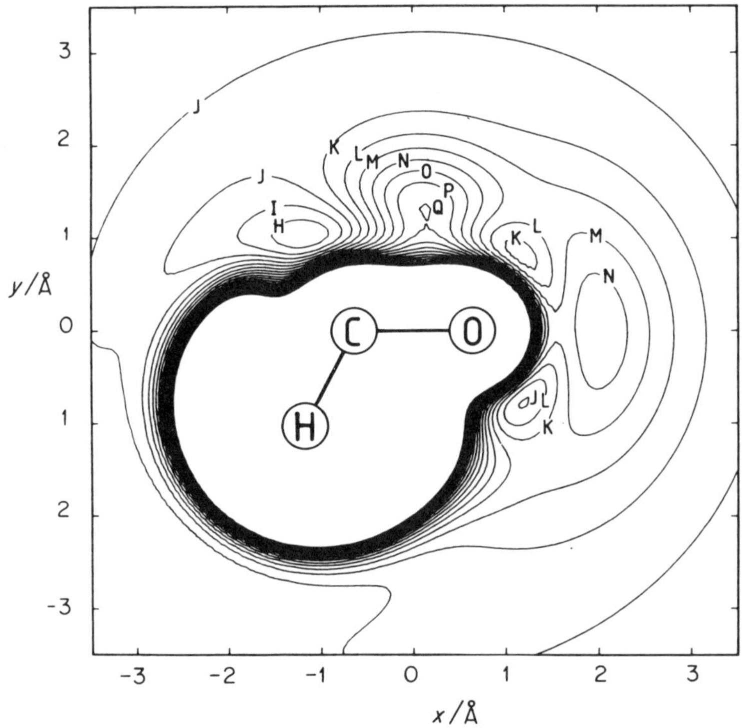

Figure 12.3 Contours of the two- plus three-body potential of H_2CO. The HCO fragment is fixed at its geometry in the equilibrium configuration and the potential is plotted for positions of the other H atom in the HCO plane. Contour H = −12.5 eV, contour interval 0.5 eV

coordinates are totally symmetric (A_1), in contrast to the four totally symmetric S_i coordinates. The difference between the two sets rests essentially in the out-of-plane valence coordinate $Q_4 = \gamma$ which has B_1 symmetry. By applying the chain rule for differentiation, it can be shown that the force constant F_{44} can determine the coefficient of the linear polynomial term in S_4. The four terms at the quadratic level which are not required to reproduce the equilibrium data are therefore S_4^2, S_1S_4, S_2S_4 and S_3S_4.

The second set of data is that pertaining to the $H_2 \cdots CO$ saddle point which we take to be the energy and geometry given in Table 12.1. As for the equilibrium structure the planarity of the saddle point reduces by one the number of independent geometrical parameters, and there are therefore six pieces of data altogether. Following arguments similar to those in Chapter 10 and Appendix 2 for multiple-minima triatomics, it can be shown that the terms S_4^2, S_1S_4, S_2S_4, S_3S_4, $S_3S_5^2$ and $S_3^2S_5^2$ are sufficient to describe the saddle point data.

The experimental data described above were used as initial input to the fitting procedure, and the range function parameters γ_i (12.2) were varied to give a potential with no spurious behaviour[12]. A vibrational variation program was

Table 12.3 Many-body expansion of the H_2CO potential[12]. The two-body terms are ER functions (1.19). The three-body terms are products of polynomials in the variables ρ_i and range functions (3.26) in the variables S_i. For the four-body term both the polynomial and range function are in S_i variables (12.2) which are the symmetry combinations of ρ_i defined in Table 12.2.

Two-body terms:

Species	D_e/eV	R_e/Å	a_1/Å^{-1}	a_2/Å^{-2}	a_3/Å^{-3}
CH	3.648	1.120	3.726	3.086	1.877
CO	11.224	1.128	3.692	1.546	1.538
OH	4.632	0.971	4.535	5.031	3.926
H_2	4.750	0.742	3.880	3.741	3.255

Three-body terms:

Parameter	HCO	CH_2	H_2O	Parameter	HCO	CH_2	H_2O
C_1/Å^{-1}	0.486	−0.854	0.496	C_{1111}/Å^{-4}	1.198	—	0.660
C_2	0.428	−0.854	0.496	C_{2222}	0.875	—	0.660
C_3	0.185	3.964	0.235	C_{3333}	0.551	—	0.237
C_{11}/Å^{-2}	−0.846	−20.613	−0.048	C_{1112}	−1.426	—	−0.316
C_{22}	0.004	−20.613	−0.048	C_{1122}	2.624	—	0.353
C_{33}	−0.503	−25.692	0.126	C_{1222}	−1.749	—	−0.316
C_{12}	0.694	−41.970	0.740	C_{1113}	−2.397	—	−0.330
C_{13}	2.364	46.835	0.415	C_{1133}	2.624	—	0.474
C_{23}	2.394	46.835	0.415	C_{1333}	−1.102	—	−0.302
C_{111}/Å^{-3}	−0.071	—	0.508	C_{2223}	−1.749	—	−0.330
C_{222}	−0.071	—	0.508	C_{2233}	2.624	—	0.474
C_{333}	−0.071	—	0.266	C_{2333}	−2.073	—	−0.302
C_{112}	−0.514	—	0.089	C_{1123}	−0.971	—	2.073
C_{122}	−0.514	—	0.089	C_{1223}	0.000	—	2.073
C_{113}	−0.514	—	−0.431	C_{1233}	0.971	—	2.074
C_{133}	−0.514	—	−0.185	V^0/eV	7.269	−0.424	8.036
C_{223}	−0.514	—	−0.431	γ_1/Å^{-1}	3.8	3.0	2.372
C_{233}	−0.514	—	−0.185	γ_2	0.0	3.0	2.372
C_{123}	3.293	—	2.260	γ_3	0.0	1.0	1.659

HCO: $S_1=(\rho_1+\rho_2+\rho_3)/\sqrt{3}$, $\rho_1=R_{CH}-1.575$ Å, $\rho_2=R_{CO}-1.248$ Å, $\rho_3=R_{OH}-1.516$ Å

CH_2: $S_1=\rho_1=R_{CH}-1.088$ Å, $S_2=\rho_2=R_{CH'}-1.088$ Å, $S_3=\rho_3=R_{HH}-2.018$ Å

H_2O: $S_1=\rho_1=R_{OH}-1.192$ Å, $S_2=\rho_2=R_{OH'}-1.192$ Å, $S_3=\rho_3=R_{HH}-1.122$ Å

Four-body term:

C_1/Å^{-1}	1.250	C_{12}/Å^{-2}	25.412	C_{355}/Å^{-3}	2.765
C_2	2.358	C_{13}	−5.390	C_{3355}/Å^{-4}	2.528
C_3	0.354	C_{23}	−7.311	V^0/eV	−3.926
C_4	−1.218	C_{14}	−11.414	γ_1/Å	4.243
C_{11}/Å^{-2}	10.412	C_{24}	−18.678	γ_2	9.0
C_{22}	30.442	C_{34}	3.104	γ_3	1.0
C_{33}	2.128	C_{56}	−1.113	γ_4	−3.536
C_{44}	5.988				
C_{55}	0.329				
C_{66}	0.377				

$\rho_1=R_{CH}-1.116$ Å, $\rho_2=R_{CH'}-1.116$ Å, $\rho_3=R_{CO}-1.208$ Å,

$\rho_4=R_{OH}-2.031$ Å, $\rho_5=R_{OH'}-2.031$ Å, $\rho_6=R_{HH}-1.898$ Å used to calculate the vibrational frequencies of H_2CO and D_2CO so that the potential could be optimized directly to the experimental vibration frequencies[12]. The optimum variational frequencies (cm^{-1}) for the fundamental bands of H_2CO were found to be as follows: (A_1) 2803 (2782), 1761 (1746), 1517 (1500); (B_1) 1150 (1167); (B_2) 2835 (2843), 1235 (1249). For D_2CO, whose frequencies were also used in the optimization, the corresponding values are (A_1) 2103 (2056), 1619 (1700), 1153 (1106); (B_1) 924 (938); (B_2) 2257 (2160), 968 (990). In both cases the experimental frequencies are quoted in parentheses[12]. The differences between experimental and calculated frequencies are larger than are obtained for triatomic molecules by this technique, which is not unexpected from both the increase in complexity of the potential and the greater dimension of the variational problem.

Figure 12.4 shows contours of the full potential for a hydrogen atom moving around the HCO fragment and a comparison with Figure 12.2 emphasizes the importance of the four-body term in the H_2CO equilibrium region. Figure 12.5 shows contours for the C_{2v} approach of H_2 to CO and illustrates the high barrier for the reaction. These figures are from reference 8, but the potential of Table 12.3 would give very similar figures.

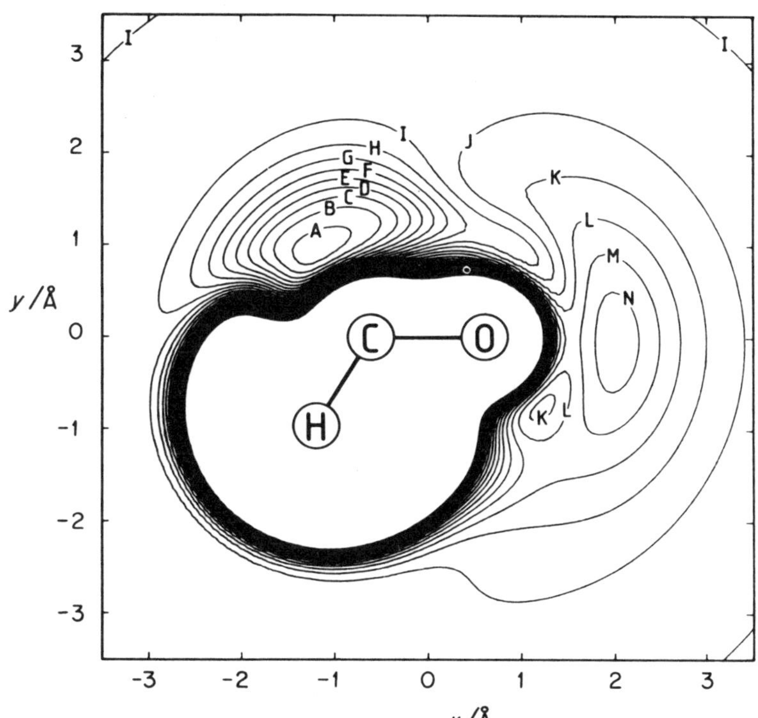

Figure 12.4 Contours of the complete H_2CO potential for the approach of H to HCO. Axes and contour labelling as in Figure 12.3. Contour A = −16.0 eV

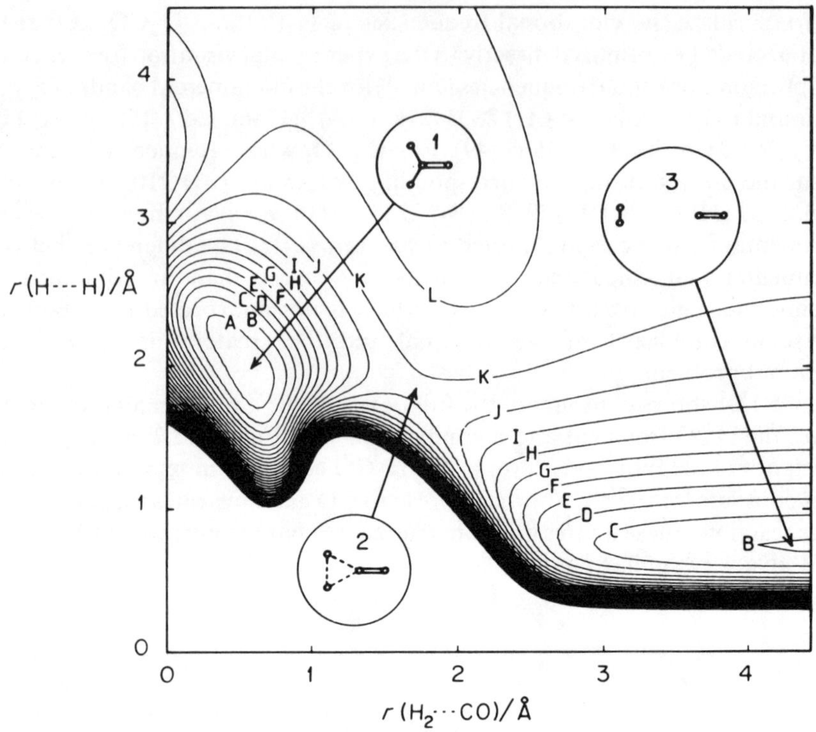

Figure 12.5 Contours showing a section of the H_2CO potential in a plane of C_{2v} symmetry in which the C–O distance is fixed at 1.21 Å. The vertical axis is the H–H distance and the horizontal axis is the distance of C to the H–H line. Structure 1 is the equilibrium configuration, 2 is the saddle point and 3 is the $H_2 + CO$ dissociation channel. Contour A = −16.0 eV, contour interval 0.5 eV

The earlier version of the H_2CO potential[8] has been used for a classical trajectory study of the reaction[13]

$$H + HCO \rightarrow H_2 + CO \qquad 12.3$$

Of particular interest is the distribution of energy in the products: most of the energy release is predicted to be in the H_2 fragment, with a substantial amount (30%) as vibrational energy. In contrast the CO fragment is predicted to be almost entirely in the ground vibrational state.

Photodissociation studies of H_2CO have been shown to produce CO in its ground vibrational state[14]. It is likely that the electronic excitation of H_2CO is followed by a rapid internal conversion to the ground electronic state and dissociation then proceeds on the ground-state surface. The trajectory calculations therefore have some relevance to the photodissociation process although the initial conditions are not, of course, the same as for reaction (12.3).

The later version of the H_2CO surface[12] has been used to calculate the classical density of rotational–vibrational states on the ground-state surface[15]. Although the surface is strongly anharmonic in the high-energy region (e.g. as the $H_2 \cdots CO$ saddle point is approached) the classical density of states was found to be fairly close to that predicted from the rigid-rotor, harmonic oscillator model. The deviations from the simple model due to anharmonicity and non-rigidity of the rotor show a fortuitous cancellation in this system and it would probably be dangerous to generalize that the simple model is always satisfactory.

12.3 The ground-state surface of H_2C_2

Like H_2CO, the ground state of H_2C_2 also has two minima on its surface but it is the metastable structure (vinylidene) which has both hydrogen atoms on the same heavy atom and the stable structure which has the hydrogen atoms on different heavy atoms. Because of the high permutational symmetry of the surface the reference structure for the four-body term must be one in which all four CH distances are equal and a planar D_{2h} structure is the most obvious choice.

The published potential functions of H_2C_2 (refs. 16, 17) have been based on the following dissociation scheme:

$$H_2C_2(\tilde{X}, {}^1\Sigma_g^+) \rightarrow \begin{array}{l} H({}^2S) + HCC(\tilde{X}, {}^2\Sigma^+) \\ C({}^3P) + CH_2(\tilde{X}, {}^3B_1) \\ C_2(X, {}^1\Sigma_g^+) + H_2(X, {}^1\Sigma_g^+) \\ 2CH(X, {}^2\Pi) \end{array} \qquad 12.4$$

There are no complications of many-valuedness as all atomic, diatomic and triatomic fragments are in their ground states. The energies of the dissociation fragments are summarized in Figure 12.6.

A detailed examination[16] of the truncated potential (i.e. ignoring the four-body term) shows that this potential gives a good representation of the equilibrium structure of acetylene, the calculated dissociation energy being 18.13 eV (observed 17.55 eV) and the bond lengths being $R_{CH} = 1.058$ Å (observed: 1.060 Å), $R_{CC} = 1.156$ Å (observed: 1.203 Å). This provides further evidence for the convergence of the many-body expansion. Figure 12.7 shows contours of this potential for a hydrogen atom moving around a fixed HCC fragment. It can be seen that the truncated potential does not reproduce the vinylidene structure, but instead gives a metastable structure with C_s symmetry whose dimensions are close to the calculated acetylene–vinylidene transition state (see Table 12.4). As with H_2CO, therefore, the four-body term is important for providing an accurate representation of the surface.

It was shown by Carter *et al.*[16] that by fitting a four-body term to the equilibrium data only (see Table 12.4), a potential for H_2C_2 was obtained which gave a satisfactory geometry and energy for the vinylidene minimum. This can be seen from Figure 12.8, which shows contours of the full potential which exactly reproduces the equilibrium data. Comparison with Figure 12.7 highlights, in particular, the improvement to the vinylidene region of the surface. However, it was later noted that this surface failed to reproduce the barrier predicted by *ab*

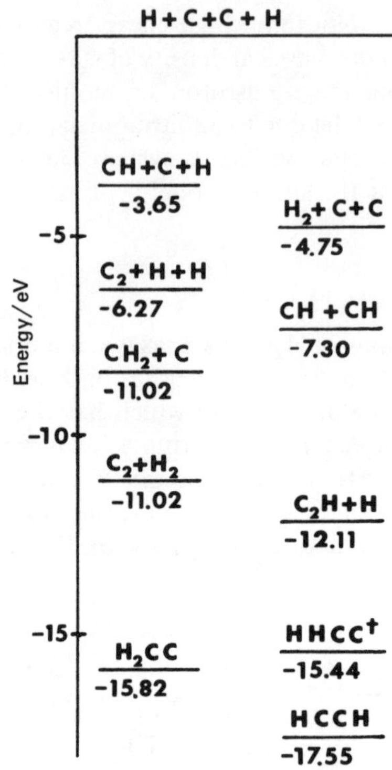

Figure 12.6 Relative energies of the dissociation fragments and the equilibrium configuration of H_2C_2 which are fitted exactly by the potential of Table 12.5. The *ab initio* energies of the vinylidene structure H_2CC and the transition state to this structure HHCC are also shown and the potential reproduces these with good accuracy

initio calculations[18] for the collinear dissociation of acetylene into two CH fragments.

In reference 17, a four-body term was constructed, in a similar way to that previously used for H_2CO, so as to reproduce all of the data given in Table 12.4. The complete potential is given in Table 12.5. Use was made of a local mode variational procedure[17] to refine the potential to the observed bond stretching vibrational frequencies. A comparison of experimental and calculated frequencies of H_2C_2 and D_2C_2 is given in Table 12.6, from which it can be seen that the equilibrium properties are well represented by the potential in Table 12.5.

In deriving the H_2C_2 potential, the γ_i values of the range function T(12.2) were optimized to give a barrier to the formation of two CH fragments, whose position and energy were closest to the *ab initio* predictions[18]. The existence of this barrier suggests that the dissociation scheme (12.3) is not completely adequate, since its origin lies in an avoided intersection of two surfaces, one dissociating to $2CH(X, {}^2\Pi)$ and the other to $2CH(a, {}^4\Sigma^-)$. As for formaldehyde, therefore, a full potential for acetylene is multi-valued, but that given in Table 12.5 will be sufficient for most uses.

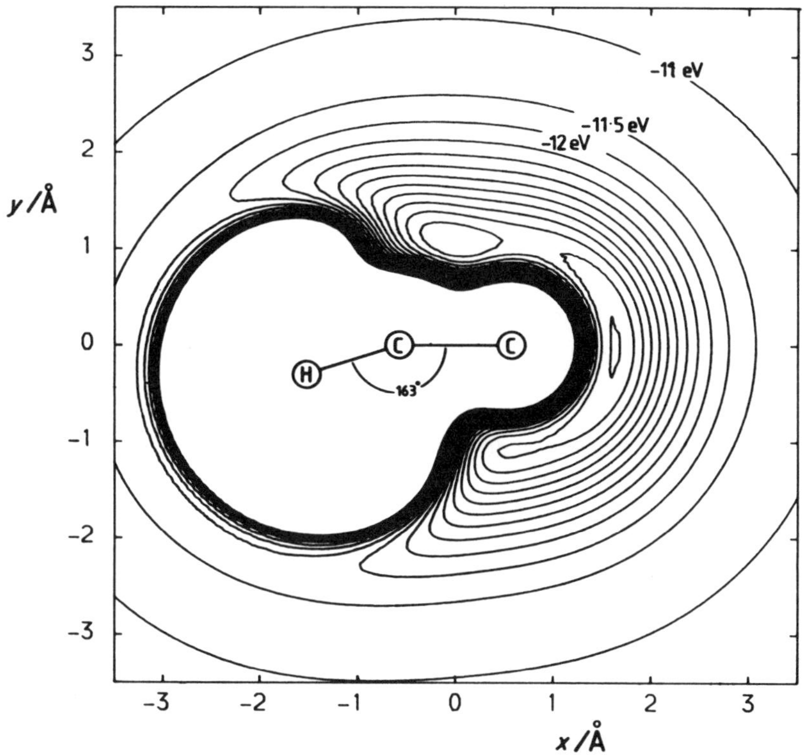

Figure 12.7 Contours of the two- plus three-body potential of H_2C_2. Contours are shown for one H moving around HCC in a geometry which is optimum for the bridging structure which corresponds to one of the minima in the figure. All four atoms are coplanar

12.4 An approximate single-valued potential for H_2O_2

In the preceding sections we have shown that the many-body expansion is a good approach to the potential functions of tetra-atomic species. However, in both examples that have been given we have had to accept that there are some regions of the surface where the functions are invalid or inaccurate because of the failure to reproduce the surface intersections in these regions. For H_2CO, the potential was not accurate in regions leading to the H_2O+C dissociation channel and for H_2C_2 the barrier for the collinear dissociation to two CH fragments was not correctly reproduced. By using a full many-valued potential (Chapter 9) these inaccuracies can be avoided for triatomic species and in principle the same can be said for tetra-atomics. However, in practice many-valued functions will usually be too difficult to derive and too complicated to use. The alternative is to use an approximate single-valued representation of the surface following the method described in Section 9.4 and we now illustrate this by an examination of the H_2O_2 surface.

Table 12.4 Properties of stationary points on the H_2C_2 surface. ΔE are energies relative to the equilibrium structure.

Equilibrium structure HCCH (refs. 9, 19)			
$r_1 = r_2 = R_{CH}/\text{Å}$	1.061	$f_{11}/\text{aJ Å}^{-2}$	6.370
$r_3 = R_{CC}/\text{Å}$	1.203	f_{33}	16.341
$\alpha = \alpha' = \text{C}\hat{\text{C}}\text{H}$	180.0°	f_{12}	−0.019
D_e/eV	17.552	f_{13}	−0.095
		$f_{\alpha\alpha}/\text{aJ}$	0.251
		$f_{\alpha\alpha'}$	0.093
Vinylidene structure H_2CC (ref. 20) (variables as in Figure 12.1a)			
$r_1 = r_2/\text{Å}$	1.096	$\beta_1 = \beta_2$	121.1°
$r_3/\text{Å}$	1.342	α	117.9°
$\Delta E/\text{eV}$	1.735	γ	0.0°
Acetylene–vinylidene transition state[20]			
$R_{CH}/\text{Å}$	1.079	$\text{H}\hat{\text{C}}\text{C}'$	176.1°
$R_{C'H'}/\text{Å}$	1.260	$\text{C}\hat{\text{C}}'\text{H}'$	67.6°
$R_{CC'}/\text{Å}$	1.289	$\Delta E/\text{eV}$	2.108
The structure is planar			

We start by writing down the complete set of adiabatic dissociation channels for H_2O_2:

$$H_2O_2(\tilde{X}, {}^1A) \rightarrow \begin{array}{ll} 2\text{OH}(X, {}^2\Pi) & \\ \text{HO}_2(\tilde{X}, {}^2A'') + \text{H}({}^2\text{S}) & \\ \left\{ \begin{array}{l} \text{O}_2(X, {}^3\Sigma_g^-) + \text{H}_2(b, {}^3\Sigma_u^+) \\ \text{O}_2(a, {}^1\Delta_g) + \text{H}_2(X, {}^1\Sigma_g^+) \end{array} \right. & \begin{array}{l} (i) \\ (ii) \end{array} \\ \left\{ \begin{array}{l} \text{H}_2\text{O}(\tilde{X}, {}^1A_1) + \text{O}({}^1\text{D}) \\ \text{H}_2\text{O}(\tilde{a}, {}^3B_1) + \text{O}({}^3\text{P}) \end{array} \right. & \begin{array}{l} (iii) \\ (iv) \end{array} \end{array} \qquad 12.5$$

The choice of channels (*i*) and (*ii*), or (*iii*) and (*iv*), is based on the usual criteria. For example, it is not possible, by spin conservation, to produce both O_2 and H_2 ground states from a singlet H_2O_2 surface. Because of this, the two channels (*i*) and (*ii*) must be invoked, channel (*i*) being reached if the H—H bond distance is long and the O—O bond distance is short, and channel (*ii*) if the converse is true. We can also see that the surface of H_2O_2 must be *at least* four-valued, since, as has been shown in Section 9.2, the ground state of $H_2O(\tilde{X}, {}^1A_1)$ is itself two-valued:

$$H_2O(\tilde{X}, {}^1A_1) \rightarrow \begin{array}{ll} \left\{ \begin{array}{l} \text{H}_2(X, {}^1\Sigma_g^+) + \text{O}({}^1\text{D}) \\ \text{H}_2(b, {}^3\Sigma_u^+) + \text{O}({}^3\text{P}) \end{array} \right. & \begin{array}{l} (v) \\ (vi) \end{array} \\ \left\{ \begin{array}{l} \text{OH}(X, {}^2\Pi) + \text{H}({}^2\text{S}) \\ \text{OH}(A, {}^2\Sigma^+) + \text{H}({}^2\text{S}) \end{array} \right. & \begin{array}{l} (vii) \\ (viii) \end{array} \end{array} \qquad 12.6$$

We can see immediately from (12.5, *iii*) and (12.6, *v* and *vi*) that there will be some

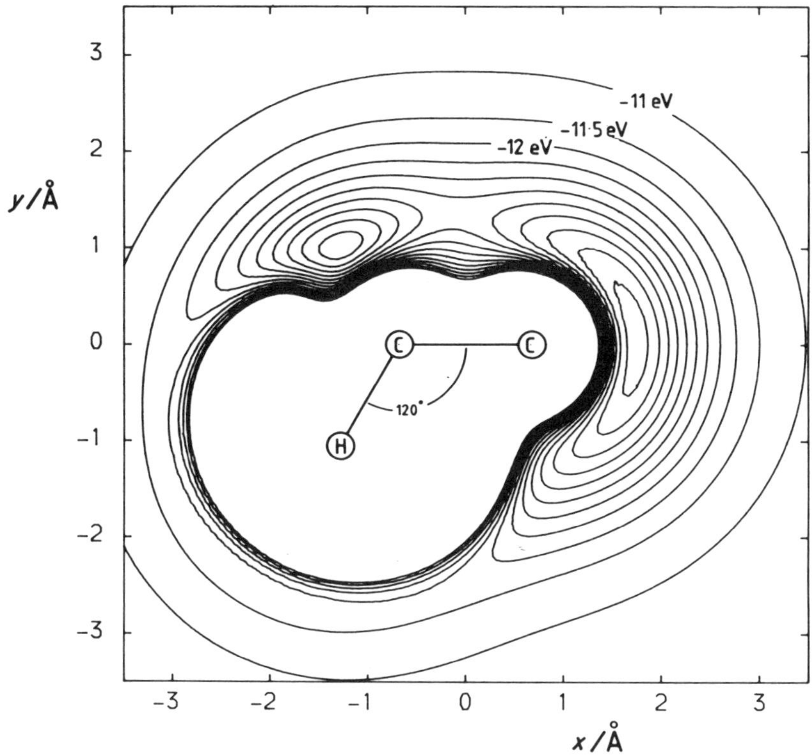

Figure 12.8 Contours of the H_2C_2 potential given in Table 12.5. The contours are for H moving around HCC in the HCC plane with the HCC geometry optimum for the vinylidene structure

regions of the H_2O_2 surface that will dissociate to two oxygen atoms both in their $O(^1D)$ states, and others that will dissociate to $O(^1D)$ and $O(^3P)$. Both O_2 states in (12.5), however, dissociate to two $O(^3P)$ atoms, and hence two further O_2 states will be necessary to cover the remaining H_2O channels.

Following the discussion in Section 9.4 we select only those channels in (12.5) which have the lower energy when the fragments are in their equilibrium configuration. This leads to the following scheme:

$$H_2O_2(\tilde{X}, {}^1A) \rightarrow \begin{cases} 2OH(X, {}^2\Pi) & (i) \\ H_2(X, {}^1\Sigma_g^+) + O_2(a, {}^1\Delta_g) & (ii) \\ H_2O(\tilde{X}, {}^1A_1) + O(^1D) & (iii) \\ HO_2(\tilde{X}, {}^2A'') + H(^2S) & (iv) \end{cases} \qquad 12.7$$

Such a surface could be used, for example, to study the dynamics of H with the ground state of HO_2 to give the products in (*i*), (*ii*) and (*iii*) of (12.7).

Of the triatomic species given in (12.7), we have already discussed H_2O in Section 9.4. We recall its assumed dissociation scheme:

Table 12.5 Many-body expansion of the H_2C_2 potential[16,17]. Format as in Table 12.3.

Two-body terms:

Species	D_e/eV	R_e/Å	a_1/Å^{-1}	a_2/Å^{-2}	a_3/Å^{-3}
C_2	6.271	1.243	5.089	6.898	4.059

H_2 and CH as in Table 12.3

Three-body terms:

C_2H:

C_1/Å^{-1}	−0.494	C_{11}/Å^{-2}	−0.781
C_2	0.842	C_{22}	−2.799
C_3	−0.494	C_{33}	−0.781
V^0/eV	1.844	C_{12}	−0.181
γ_1/Å^{-1}	2.0	C_{13}	1.937
$\gamma_2=\gamma_3$	0.0	C_{23}	−0.181

$S_1=(\rho_1+\rho_2+\rho_3)/\sqrt{3},\ \ \rho_1=R_{CH}-1.551\ \text{Å},\ \ \rho_2=R_{CC}-1.511\ \text{Å},\ \ \rho_3=R_{CH'}-1.511\ \text{Å}$

CH_2 as in Table 12.3

Four-body term:

C_1/Å^{-1}	0.1111	C_{166}/Å^{-3}	−0.0187
C_2	−1.3197	C_{266}	−0.0814
C_3	−1.7541	C_{344}	0.3392
C_{11}/Å^{-2}	0.4075	C_{355}	−0.1101
C_{22}	12.1652	C_{366}	0.0981
C_{33}	−0.5569	C_{6666}/Å^{-4}	0.0578
C_{44}	−0.1525	V^0/eV	4.4124
C_{55}	0.0467	γ_1/Å^{-1}	2.9
C_{66}	−0.5076	γ_2	−9.0
C_{12}	−4.2566	γ_3	−1.5
C_{13}	−0.3505		
C_{23}	3.3245		

$\rho_i\ (i=1,2,4,5)=R_{CH}-1.5000\ \text{Å},\ \ \rho_3=S_2=R_{CC}-1.2033\ \text{Å},\ \ \rho_6=S_3=R_{HH}-2.7481\ \text{Å}$

$S_1=\rho_1+\rho_2+\rho_4+\rho_5,\ \ S_4=\rho_1+\rho_2-\rho_4-\rho_5,\ \ S_5=\rho_1-\rho_2+\rho_4-\rho_5,$
$S_6=\rho_1-\rho_2-\rho_4+\rho_5$

$$H_2O(\tilde{X},{}^1A_1)\rightarrow \begin{matrix} H_2(X,{}^1\Sigma_g^+)+O(^1D) \\ OH(X,{}^2\Pi)+H(^2S) \end{matrix} \qquad 12.8$$

The remaining triatomic, HO_2, presents no problems as it dissociates to ground state fragments at all asymptotic limits according to the scheme

$$HO_2(\tilde{X},{}^2A'')\rightarrow \begin{matrix} O_2(X,{}^3\Sigma_g^-)+H(^2S) \\ OH(X,{}^2\Pi)+O(^3P) \end{matrix} \qquad 12.9$$

The configuration of the H_2O_2 species is defined by values of the six internuclear distances which we label as in Figure 12.9. The two H_2O potentials will be $V_{HHO}(R_1, R_5, R_6)$, already written as (9.20), and an equivalent function

Table 12.6 Comparison of experimental and calculated bond stretching frequencies (cm^{-1}) for acetylene[17]. ν_1 = sym. CH stretch, ν_2 = CC stretch, ν_3 = asym. CH stretch.

	H_2C_2		D_2C_2	
Assignment	Exptl.	Calc.	Exptl.	Calc.
100	3373	3372	2705	2686
001	3295	3295	2439	2448
002	6502	6510	—	—
101	6556	6536	5097	5069
200	6709	6700	—	—
003	9640	9616	7230	7239
201	9835	9792	7734	7685
010	1974	1971	1765	1783
011	5260	5250	4191	4212
111	8512	8495	6829	6826

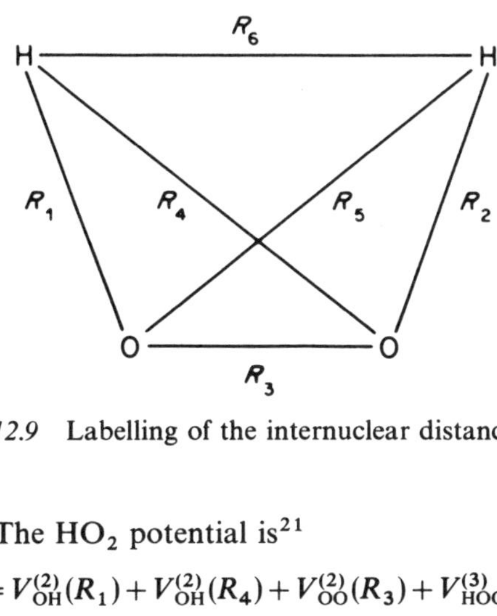

Figure 12.9 Labelling of the internuclear distances for H_2O_2

$V_{HHO}(R_2, R_4, R_6)$. The HO_2 potential is[21]

$$V_{HOO}(R_1, R_3, R_4) = V^{(2)}_{OH}(R_1) + V^{(2)}_{OH}(R_4) + V^{(2)}_{OO}(R_3) + V^{(3)}_{HOO}(R_1, R_3, R_4) \quad (12.10)$$

where the two-body terms are for the diatomic states contained in (12.9). There is an equivalent function $V_{HOO}(R_2, R_3, R_5)$.

We note that the only two-body term for the OH species that appears in (12.8) and (12.9) is for the $^2\Pi$ state and that this is the same state that occurs in the H_2O_2 dissociation scheme (12.7). Likewise, the only H_2 state that appears in (12.8) and satisfies (12.7, *ii*) is $^1\Sigma_g^+$. Thus these two-body terms present no ambiguity in the H_2O_2 potential. For O_2, however, we have two conditions to satisfy. The two-body potential must represent the $^3\Sigma_g^-$ state in (12.10) but the $^1\Delta_g$ state in the limit (12.7, *ii*). To achieve this we write the O_2 contribution to the H_2O_2 potential in the form

$$gV^{(2)}_{OO(a)}(R_3) + (1-g)V^{(2)}_{OO(b)}(R_3) \quad 12.11$$

where (a) and (b) label the $^3\Sigma_g^-$ and $^1\Delta_g$ potentials, respectively, and g is a function of the variables R_i which takes the limits 0 or 1 in the asymptotic regions as appropriate.

In all but three asymptotic regions both $V^{(2)}_{\mathrm{OO}(a)}$ and $V^{(2)}_{\mathrm{OO}(b)}$ are zero; hence we need to concern ourselves only with the behaviour of g in these three. In one of them [scheme (12.7, *ii*)] $R_1 = R_2 = R_4 = R_5 = \infty$, and we require g to approach zero in this limit. In the other two [scheme (12.7, *iv*)] we have either $R_2 = R_5 = R_6 = \infty$ or $R_1 = R_4 = R_6 = \infty$ and in these g must approach unity. Using the variable

$$y = \rho_1 + \rho_2 + \rho_4 + \rho_5 - 4\rho_6, \quad \rho_i = R_i - R_i^0 \tag{12.12}$$

the function

$$g = \tfrac{1}{2}[1 - \tanh(\beta y/2)] \tag{12.13}$$

satisfies these requirements.

The one-body term, being the energy of the oxygen atom, must also be associated with a switching function. The O(^{1}D) energy must only appear in the dissociation limit (12.7, *iii*) which is when $R_1 = R_3 = R_5 = \infty$ and $R_2 = R_3 = R_4 = \infty$, but it must in these limits appear both for the free oxygen atom and for the oxygen atom in the H_2O fragment; in the latter it must have the same form as the switching function f in (9.20). We therefore write the oxygen atom component of the H_2O_2 potential as

$$V^{(1)}_{\mathrm{O}}[f(R_1, R_5, R_6) + f(R_2, R_4, R_6)]h(R_i) \tag{12.14}$$

which, when $R_2 = R_3 = R_4 = \infty$, becomes

$$V^{(1)}_{\mathrm{O}}[f(R_1, R_5, R_6) + 1] \tag{12.15}$$

and when $R_1 = R_3 = R_5 = \infty$ becomes

$$V^{(1)}_{\mathrm{O}}[1 + f(R_2, R_4, R_6)] \tag{12.16}$$

and in all other limits is zero. The function

$$h = \tfrac{1}{2}[1 - \tanh[\delta(\rho_1 + \rho_2 + \rho_4 + \rho_5 - 4\rho_3)/2]\} \tag{12.17}$$

satisfies these requirements.

The complete single-valued representation of the H_2O_2 potential is therefore

$$\begin{aligned} V_{\mathrm{HHOO}} = {} & V^{(1)}_{\mathrm{O}}[f(R_1, R_5, R_6) + f(R_2, R_4, R_6)]h + V^{(2)}_{\mathrm{OH}}(R_1) + V^{(2)}_{\mathrm{OH}}(R_2) \\ & + V^{(2)}_{\mathrm{OH}}(R_4) + V^{(2)}_{\mathrm{OH}}(R_5) + V^{(2)}_{\mathrm{HH}}(R_6) + V^{(2)}_{\mathrm{OO}(a)}(R_3)g + V^{(2)}_{\mathrm{OO}(b)}(R_3)(1-g) \\ & + V^{(3)}_{\mathrm{HHO}}(R_1, R_5, R_6) + V^{(3)}_{\mathrm{HHO}}(R_2, R_4, R_6) + V^{(3)}_{\mathrm{HOO}}(R_1, R_3, R_4) \\ & + V^{(3)}_{\mathrm{HOO}}(R_2, R_3, R_5) + V^{(4)}_{\mathrm{HHOO}} \end{aligned} \tag{12.18}$$

where f, g and h have been defined earlier. This function has been derived by Murrell and Carter[22], but the optimum values of the parameters have not yet been calculated. The task of producing such a potential is certainly greater than for the H_2CO and H_2C_2 potentials discussed earlier but we believe that with adequate data on the stationary points on the potential surface, particularly the equilibrium configuration and a (possible) metastable H_2OO, the task can be completed. In contrast, we do not believe that the full four-valued surface of H_2O_2 which is necessary to satisfy all the limits defined in (12.5) could at present be deduced.

References

1. M. Karplus, L. Pedersen and K. Morokuma, *J. Am. Chem. Soc.*, **89**, 5064 (1964).
2. L. M. Raff, L. Stivers, R. N. Porter, D. L. Thompson and L. B. Sims, *J. Chem. Phys.*, **52**, 3449 (1970).
3. J. C. Tully, 'Diatomics in Molecules', Ch. 6 in *Modern Theoretical Chemistry*, Vol. 7A, *Semiempirical Methods of Electronic Structure Calculation*, Ed. G. A. Segal, Plenum Press, New York, 1977.
4. P. J. Kuntz, 'Semiempirical Potential Energy Surfaces', in *The Theory of Chemical Reaction Dynamics*, Ed. M. Baer, CRC Press, Cleveland, Ohio, 1984.
5. C. W. Eaker and C. A. Parr, *J. Chem. Phys.*, **65**, 5155 (1976).
6. M. J. H. Kemper, J. M. F. Van Dijk and H. M. Buck, *J. Am. Chem. Soc.*, **100**, 7841 (1978).
7. R. L. Jaffe, D. M. Hayes and K. Morokuma, *J. Chem. Phys.*, **60**, 5108 (1974).
8. S. Carter, I. M. Mills and J. N. Murrell, *Mol. Phys.*, **39**, 455 (1980).
9. JANAF, *Tables of Thermochemical Data*, NBS Technical Note, 2nd ed., National Bureau of Standards, Washington, DC, 1971.
10. K. Takagi and T. Oka, *J. Phys. Soc. Jpn.*, **18**, 1174 (1963).
11. J. L. Duncan and P. D. Mallinson, *Chem. Phys. Lett.*, **23**, 597 (1973).
12. N. C. Handy and S. Carter, *Chem. Phys. Lett.*, **79**, 118 (1981); **83**, 216 (1981).
13. S. C. Farantos and J. N. Murrell, *Mol. Phys.*, **40**, 883 (1980).
14. P. L. Houston and C. B. Moore, *J. Chem. Phys.*, **65**, 757 (1976).
15. S. C. Farantos, J. N. Murrell and J. C. Hajduk, *Chem. Phys.*, **68**, 109 (1982).
16. S. Carter, I. M. Mills and J. N. Murrell, *Mol. Phys.*, **41**, 191 (1980).
17. L. Halonen, S. Carter and M. S. Child, *Mol. Phys.*, **47**, 1097 (1982).
18. M. Raimondi, M. Simonetta and J. Gerratt, *Chem. Phys. Lett.*, **77**, 12 (1981).
19. G. Strey and I. M. Mills, *J. Mol. Spectrosc.*, **59**, 103 (1976).
20. C. E. Dykstra and H. F. Schaefer, *J. Am. Chem. Soc.*, **100**, 1378 (1978).
21. S. Farantos, E. C. Leisegang, J. N. Murrell, K. Sorbie, J. J. C. Texeira-Dias and A. J. C. Varandas, *Mol. Phys.*, **34**, 947 (1977).
22. J. N. Murrell and S. Carter, *J. Phys. Chem.*, in press.

DIP MY TOE IN THAT, LORD? YOU MUST BE JOKING.
DO YOU MEAN TO TELL ME THAT ALL I HAVE TO DO TO GET RID OF AN INCURABLE DISEASE, IS TO DIP MY TOE IN A SMALL PUDDLE?
WHY! A CHILD OF THREE COULD DO THAT.
WHO DO YOU THINK I AM?
I QUITE EXPECTED TO CLIMB MOUNT EVEREST. OR LEAD AN EXPEDITION TO THE NORTH POLE
OR WRITE A BOOK ON THEOLOGY, OR AT LEAST DO SOMETHING BRAVE OR DRAMATIC OR EXTRAORDINARY

Figure 13.1 Reproduced by kind permission of the Rev. Graham Jeffery. Published in *The Gospel According to Barnabas*, Mowbrays, London and Oxford, 1975

Chapter 13

Conclusion

We hope that the message of our cartoon evokes a sympathetic response from our readers. We show it to emphasize that an important principle of the scientific method is that one should start with the simple and introduce complexity in stages when the simple is shown to be inadequate. That is how diatomic potentials have been attacked; the simple functions such as Morse, Rydberg, Born–Mayer, etc., give a qualitatively correct description of the potential and only modest extensions are needed to obtain functions which stand up to the most stringent experimental tests.

In this book we have tried to show that the potential functions of tri- and tetra-atomic species can be developed along the same lines as diatomic functions. We are not, of course, the first to take this path, as the history of the LEPS function (Section 3.2) has been from the simple to the less simple. However, LEPS functions were designed for surfaces with the H_3 topography and they do not have the flexibility to describe all types of surface.

The many-body expansion which we have favoured in this book is not in itself a simplification of the potential function but it does provide a strategy for building up the function from the potentials of the dissociation fragments. By imposing the correct dissociation limits to a potential one at least obtains a function which may be useful for scattering calculations; without the correct limits the function will almost certainly not be useful. Moreover, as was shown in Chapter 7, it is sometimes possible to extrapolate from regions of the surface where the potential may be deduced from spectroscopy to the dissociation limits. Morse or Rydberg functions are, overall, better diatomic potentials than the Taylor expansion about the minimum because they have the correct asymptotic limit. Similarly, the dissociation limits of a polyatomic potential provide an important constraint for the whole function.

There are two other important features of the many-body expansion. Firstly, it provides a simple method for making empirical corrections to an *ab initio* surface. The one- and two-body terms are usually known exactly from experiment and combining these with an *ab initio* three-body term should provide a very good surface. The second feature is that for many-valued surfaces the off-diagonal elements can be restricted to three-body terms and higher orders.

Examination of the triatomic potentials which have been given in this book probably provoke the response that they are not *that* simple. We think that this response should be tempered by two inescapable factors: firstly that they have a high dimensionality and secondly that to treat surface intersections correctly requires the use of many-valued functions. Given these two factors, we believe that the many-body functions are no more complicated than might be expected. If one requires n parameters to represent accurately a diatomic potential then one can expect n^3 parameters to be needed for a triatomic potential with the same level of accuracy and $3n^3$ to be needed for a two-valued surface (two diagonal and one off-diagonal component of the potential matrix).

The functions used to represent the three-body terms in this book have generally been taken as the product of a polynomial and an exponentially decaying function based on the hyperbolic tangent. In our experience this has proved very robust as it has usually led to a satisfactory surface. The greatest difficulties have been encountered in trying to fit the parameters of the three-body term to the spectroscopic constants (Section 7.1) of multi-minima surfaces. The choice of polynomial terms in such cases is not always unique, notwithstanding the strategy summarized in Appendix 2, and as one is fitting the potential only to data in selected regions of the surface there are no quantitative restraints on the function in other regions. Our experience has been that fitting by a least-squares criterion to *ab initio* points distributed over the whole surface, using a complete quartic polynomial in the three-body term, is a much easier task. We would encourage an open mind on the functional forms used for the many-body terms. The best functions may have yet to be found.

For systems with more than three atoms, an important question is whether the many-body expansion is convergent. We should not be too general in our enquiry. The relevant question is whether there is convergence over those regions of the surface which are important for the particular dynamical problem in mind. Chemists, for example, will generally only be interested in those regions of the surface which are important for vibrational spectroscopy and low-energy collisions, and the many-body expansion is more likely to be convergent for these than for high-energy regions.

Convergence of the many-body expansion does not mean that three- and four-body terms can be ignored. The only case where this is likely to be an acceptable simplification is when there is at least one inert gas atom in the cluster. However, our experience with tetra-atomic surfaces suggests that it is possible to fit the four-body term by a relatively simple function and obtain a reasonable fit to the surface. The main features of the surface are contained in the one-, two- and three-body terms and the main role of the four-body term is to reproduce the correct relative energies of stable and metastable structures.

The only systems with more than four atoms for which the many-body expansion has been examined in detail are H_5 and H_6 (ref. 1). Taking the two-, three- and four-body terms from *ab initio* calculations on H_2, H_3 and H_4, and neglecting higher order terms, Varandas and Murrell predicted bond lengths and energies for the optimum D_{5h} form of H_5 and D_{6h} form of H_6. Comparison with *ab initio* calculations on these species showed that the bond lengths were predicted accurately but the energies were too low. It appears that the combined five- and six-body energies in H_6 might be approximately 200 kJ mol^{-1}, which by chemical standards is large. However, this conclusion might be changed if the H_n surfaces were re-examined using more recent *ab initio* data and a full many-valued representation of the surface. Moreover, the H_n regular polygons are delocalized electron structures (H_6 is similar to the π-electron component of benzene) and high-order terms might be expected to be larger than usual. In summary, we take at present an uncommitted view of the convergence of the many-body expansion for systems with more than four atoms.

Finally, we turn to the question of how best to disseminate information on potential energy functions. We believe that there is a strong case to be made for a computerized database, particularly if this is networked to provide remote access and file transfer. The following points justify this view.

1. The functions are generally complicated and the possibility of transposition errors between written records is high.
2. The functions will need to be frequently updated in the light of new experimental or computational information or a better fitting to old data. Such updating may not bring forward new points of principle which would merit publication in a scientific journal, but the changes to a potential may have a significant effect on dynamical calculations.
3. The examination of multiparameter functions is greatly enhanced by graphical display and for multidimensional functions this facility must provide for a choice of cross-sections through the surface.

Before a computer database can be established which will satisfy the above requirements it is necessary to have a limited set of function forms which represent the potentials. These functions are written into the operating program and the parameters of the potential functions are stored as data. We have established two databases, DIATOMS and TRIATOMS, on a PRIME 550 computer which is networked throughout the UK[2]. The programs are 'user-friendly' and robust, and experience has shown that they can be used immediately by anyone with moderate scientific knowledge in the field.

DIATOMS contains spectroscopic data on diatomic states such as the equilibrium bond length, dissociation energy and force constants. RKR or *ab initio* data are also included. All data are referenced. The potentials are represented by the Extended Rydberg function (1.19) and its extension to van der Waals systems (5.9). The data can be listed on command or displayed graphically in any range of internuclear distances. DIATOMS provides the two-body terms needed for the many-body expansion which is used for TRIATOMS.

TRIATOMS contains spectroscopic data pertaining to equilibrium structures on the surface and *ab initio* data which are particularly relevant to the surface. Parameters of the many-body expansion are recorded for both single- and two-valued functions. Parameters are also given for the approximate single-valued representations of two-valued surfaces using the method described in Section 9.4.

There is a graphical package that provides for one-dimensional cross-sections (two variables held constant and the potential plotted as a function of the third) or as contours of a two-dimensional cross-section. There are a number of choices for the two-dimensional cross-sections, some more suitable for spectroscopic analysis such as bond length or bond angle variation, and others more suitable for scattering studies, e.g. using the variables of Figure 4.1b. Most of the figures given in this book were obtained from the TRIATOMS graphical package.

A computer program POT 3 has been written which incorporates all methods of obtaining the many-body expansion of a triatomic potential which have been described in this book. Programs are also available for carrying out the variational calculation of vibration frequencies from such potentials. All programs are in Standard Fortran 4. Enquiries regarding access to these programs should be addressed to Dr. S. Carter, Department of Chemistry, The University of Reading, Reading, RG6 2AD, U.K.

References

1. A. J. C. Varandas and J. N. Murrell, *Faraday Discuss., Chem. Soc.*, **62**, 92 (1977).
2. S. Carter, P. Huxley and J. N. Murrell, *Chem. Brit.*, **18**, 431 (1982).

Appendix 1

Relationship between the derivatives of the potential in the coordinate sets (r_1, r_2, α) and (R_1, R_2, R_3)

For convenience, we functionally distinguish the potential in the two coordinate sets

$$V(R_1, R_2, R_3) \equiv W(r_1, r_2, \alpha)$$

where

$$R_1 \equiv r_1$$

$$R_2 \equiv r_2$$

$$R_3^2 = r_1^2 + r_2^2 - 2r_1 r_2 \cos \alpha$$

The required transformations are given below using the notation

$$V_{12} = \frac{\partial^2 V}{\partial R_1\, \partial R_2}, \quad W_{1\alpha} = \frac{\partial^2 W}{\partial r_1\, \partial \alpha}, \quad R3_{1\alpha} = \frac{\partial^2 R_3}{\partial r_1\, \partial \alpha}, \quad \text{etc.}$$

Permuting 1 and 2 gives further expressions. The derivatives of W with respect to r_1, r_2 and α correspond to the spectroscopic force constants; $f_{\alpha\alpha} \equiv W_{\alpha\alpha}$, etc.

Linear

$$W_1 = V_1 + V_3 R3_1$$
$$W_\alpha = V_3 R3_\alpha$$

Quadratic

$$W_{11} = V_{11} + 2V_{13}R3_1 + V_{33}R3_1^2 + V_3 R3_{11}$$
$$W_{\alpha\alpha} = V_{33}R3_\alpha^2 + V_3 R3_{\alpha\alpha}$$
$$W_{12} = V_{12} + V_{13}R3_2 + V_{23}R3_1 + V_{33}R3_1 R3_2 + V_3 R3_{12}$$
$$W_{1\alpha} = V_{13}R3_\alpha + V_{33}R3_1 R3_\alpha + V_3 R3_{1\alpha}$$

Cubic

$$W_{111} = V_{111} + 3V_{113}R3_1 + 3V_{133}R3_1^2 + V_{333}R3_1^3 + 3V_{13}R3_{11} + 3V_{33}R3_1 R3_{11} + V_3 R3_{111}$$
$$W_{\alpha\alpha\alpha} = V_{333}R3_\alpha^3 + 3V_{33}R3_\alpha R3_{\alpha\alpha} + V_3 R3_{\alpha\alpha\alpha}$$
$$W_{112} = V_{112} + V_{113}R3_2 + 2V_{123}R3_1 + 2V_{133}R3_1 R3_2 + V_{233}R3_1^2 + V_{333}R3_1^2 R3_2 + 2V_{13}R3_{12} + V_{23}R3_{11} + V_{33}(2R3_1 R3_{12} + R3_2 R3_{11}) + V_3 R3_{112}$$

$$\begin{aligned}
W_{11\alpha} = {} & V_{113}R3_\alpha + 2V_{133}R3_1R3_\alpha + V_{333}R3_1^2R3_\alpha + 2V_{13}R3_{1\alpha} + V_{33}(2R3_1R3_{1\alpha} \\
& + R3_{11}R3_\alpha) + V_3R3_{11\alpha} \\
W_{1\alpha\alpha} = {} & V_{133}R3_\alpha^2 + V_{333}R3_1R3_\alpha^2 + V_{13}R3_{\alpha\alpha} + V_{33}(2R3_\alpha R3_{1\alpha} + R3_1R3_{\alpha\alpha}) + V_3R3_{1\alpha\alpha} \\
W_{12\alpha} = {} & V_{123}R3_\alpha + R3_\alpha(V_{133}R3_2 + V_{233}R3_1) + V_{333}R3_1R3_2R3_\alpha + V_{13}R3_{2\alpha} \\
& + V_{23}R3_{1\alpha} + V_{33}(R3_1R3_{2\alpha} + R3_2R3_{1\alpha} + R3_\alpha R3_{12}) + V_3R3_{12\alpha}
\end{aligned}$$

Quartic

$$\begin{aligned}
W_{1111} = {} & V_{1111} + 4V_{1113}R3_1 + 6V_{1133}R3_1^2 + 4V_{1333}R3_1^3 + V_{3333}R3_1^4 + 6V_{113}R3_{11} \\
& + 12V_{133}R3_1R3_{11} + 6V_{333}R3_1^2R3_{11} + 4V_{13}R3_{111} \\
& + V_{33}(4R3_1R3_{111} + 3R3_{11}^2) + V_3R3_{1111} \\
W_{\alpha\alpha\alpha\alpha} = {} & V_{3333}R3_\alpha^4 + 6V_{333}R3_\alpha^2R3_{\alpha\alpha} + V_{33}(4R3_\alpha R3_{\alpha\alpha\alpha} + 3R3_{\alpha\alpha}^2) + V_3R3_{\alpha\alpha\alpha\alpha} \\
W_{1112} = {} & V_{1112} + V_{1113}R3_2 + 3V_{1123}R3_1 + 3V_{1133}R3_1R3_2 + 3V_{1233}R3_1^2 \\
& + 3V_{1333}R3_1^2R3_2 + V_{2333}R3_1^3 + V_{3333}R3_1^3R3_2 + 3V_{113}R3_{12} \\
& + 3V_{233}R3_1R3_{11} + 3V_{123}R3_{11} + 3V_{133}R3_2R3_{11} + 6V_{133}R3_1R3_{12} \\
& + 3V_{333}(R3_1R3_2R3_{11} + R3_1^2R3_{12}) + 3V_{13}R3_{112} + V_{23}R3_{111} \\
& + V_{33}(3R3_{11}R3_{12} + 3R3_1R3_{112} + R3_2R3_{111}) + V_3R3_{1112} \\
W_{1122} = {} & V_{1122} + 2V_{1123}R3_2 + 2V_{1223}R3_1 + V_{1133}R3_2^2 + V_{2233}R3_1^2 \\
& + 4V_{1233}R3_1R3_2 + 2V_{1333}R3_1R3_2^2 + 2V_{2333}R3_1^2R3_2 + V_{3333}R3_1^2R3_2^2 \\
& + V_{113}R3_{22} + V_{223}R3_{11} + 4V_{123}R3_{12} + V_{133}(4R3_2R3_{12} + 2R3_1R3_{22}) \\
& + V_{233}(4R3_1R3_{12} + 2R3_2R3_{11}) + V_{333}(4R3_1R3_2R3_{12} + R3_{22}R3_1^2 \\
& + R3_{11}R3_2^2) + 2V_{13}R3_{122} + 2V_{23}R3_{112} + V_{33}(2R3_{12}^2 + 2R3_1R3_{122} \\
& + 2R3_2R3_{112} + R3_{11}R3_{22}) + V_3R3_{1122} \\
W_{111\alpha} = {} & V_{1113}R3_\alpha + 3V_{1133}R3_1R3_\alpha + 3V_{1333}R3_1^2R3_\alpha + V_{3333}R3_1^3R3_\alpha \\
& + 3V_{113}R3_{1\alpha} + V_{133}(3R3_{11}R3_\alpha + 6R3_1R3_{1\alpha}) + 3V_{333}(R3_1R3_\alpha R3_{11} \\
& + R3_1^2R3_{1\alpha}) + 3V_{13}R3_{11\alpha} + V_{33}(3R3_{11}R3_{1\alpha} + 3R3_1R3_{11\alpha} + R3_{111}R3_\alpha) \\
& + V_3R3_{111\alpha} \\
W_{11\alpha\alpha} = {} & V_{1133}R3_\alpha^2 + 2V_{1333}R3_1R3_\alpha^2 + V_{3333}R3_1^2R3_\alpha^2 + V_{113}R3_{\alpha\alpha} \\
& + V_{133}(4R3_\alpha R3_{1\alpha} + 2R3_1R3_{\alpha\alpha}) + V_{333}(R3_{11}R3_\alpha^2 + R3_{\alpha\alpha}R3_1^2 \\
& + 4R3_1R3_\alpha R3_{1\alpha}) + 2V_{13}R3_{1\alpha\alpha} + V_{33}(2R3_{1\alpha}^2 + 2R3_1R3_{1\alpha\alpha} \\
& + 2R3_\alpha R3_{11\alpha} + R3_{11}R3_{\alpha\alpha}) + V_3R3_{11\alpha\alpha} \\
W_{1\alpha\alpha\alpha} = {} & V_{1333}R3_\alpha^3 + V_{3333}R3_1R3_\alpha^3 + 3V_{133}R3_\alpha R3_{\alpha\alpha} + 3V_{333}(R3_1R3_\alpha R3_{\alpha\alpha} \\
& + R3_{1\alpha}R3_\alpha^2) + V_{13}R3_{\alpha\alpha\alpha} + V_{33}(R3_1R3_{\alpha\alpha\alpha} + 3R3_\alpha R3_{1\alpha\alpha} + 3R3_{1\alpha}R3_{\alpha\alpha}) \\
& + V_3R3_{1\alpha\alpha} \\
W_{112\alpha} = {} & V_{1123}R3_\alpha + V_{1133}R3_2R3_\alpha + 2V_{1333}R3_1R3_2R3_\alpha + V_{2333}R3_1^2R3_\alpha \\
& + 2V_{1233}R3_1R3_\alpha + V_{3333}R3_1^2R3_2R3_\alpha + 2V_{133}(R3_{12}R3_\alpha + R3_{1\alpha}R3_2 \\
& + R3_{2\alpha}R3_1) + V_{233}(R3_{11}R3_\alpha + 2R3_1R3_{1\alpha}) + V_{113}R3_{2\alpha} + 2V_{123}R3_{1\alpha} \\
& + V_{333}(2R3_1R3_\alpha R3_{12} + R3_2R3_\alpha R3_{11} + 2R3_1R3_2R3_{1\alpha} + R3_{2\alpha}R3_1^2) \\
& + 2V_{13}R3_{12\alpha} + V_{23}R3_{11\alpha} + V_{33}(R3_\alpha R3_{112} + 2R3_{1\alpha}R3_{12} + 2R3_1R3_{12\alpha} \\
& + R3_{2\alpha}R3_{11} + R3_2R3_{11\alpha}) + V_3R3_{112\alpha} \\
W_{12\alpha\alpha} = {} & V_{1233}R3_\alpha^2 + V_{3333}R3_1R3_2R3_\alpha^2 + V_{1333}R3_2R3_\alpha^2 + V_{2333}R3_1R3_\alpha^2 \\
& + V_{1233}R3_\alpha^2 + V_{333}(R3_1R3_2R3_{\alpha\alpha} + 2R3_2R3_\alpha R3_{1\alpha} + 2R3_1R3_\alpha R3_{2\alpha} \\
& + R3_{12}R3_\alpha^2) + V_{133}(R3_2R3_{\alpha\alpha} + 2R3_{2\alpha}R3_\alpha) + V_{233}(R3_1R3_{\alpha\alpha} \\
& + 2R3_{1\alpha}R3_\alpha) + V_{123}R3_{\alpha\alpha} + V_{33}(R3_{\alpha\alpha}R3_{12} + 2R3_\alpha R3_{12\alpha} + 2R3_{1\alpha}R3_{2\alpha} \\
& + R3_2R3_{1\alpha\alpha} + R3_1R3_{2\alpha\alpha}) + V_{13}R3_{2\alpha\alpha} + V_{23}R3_{1\alpha\alpha} + V_3R3_{12\alpha\alpha}
\end{aligned}$$

The derivatives $\left(\frac{\partial R_3}{\partial r_1}\right),\left(\frac{\partial^2 R_3}{\partial \alpha^2}\right)$, etc. are obtained from the following expressions.

Linear

$$R3_1=(r_1-r_2\cos\alpha)/R_3=(R_1-R_2\cos\alpha)/R_3$$
$$R3_\alpha=R_1R_2\sin\alpha/R_3$$

Quadratic

$$R3_{11}=(1-R3_1^2)/R_3$$
$$R_{3\alpha\alpha}=(R_1R_2\cos\alpha-R3_\alpha^2)/R_3$$
$$R3_{12}=-(\cos\alpha+R3_1R3_2)/R_3$$
$$R3_{1\alpha}=(R_2\sin\alpha-R3_1R3_\alpha)/R_3$$

Cubic

$$R3_{111}=-2R3_1R3_{11}/R_3-R3_1(1+R3_1^2)/R_3^2$$
$$R3_{\alpha\alpha\alpha}=-(R_1R_2\sin\alpha+2R3_\alpha R3_{\alpha\alpha})/R_3-R3_\alpha(R_1R_2\cos\alpha-R3_\alpha^2)/R_3^2$$
$$R3_{112}=-2R3_1R3_{12}/R_3-R3_2(1+R3_1^2)/R_3^2$$
$$R3_{11\alpha}=-2R3_1R3_{1\alpha}/R_3-R3_\alpha(1+R3_1^2)/R_3^2$$
$$R3_{1\alpha\alpha}=(R_2\cos\alpha-R3_1R3_{\alpha\alpha}-R3_\alpha R3_{1\alpha})/R_3-R3_\alpha(R_2\sin\alpha-R3_1R3_\alpha)/R_3^2$$
$$R3_{12\alpha}=(\sin\alpha-R3_1R3_{2\alpha}-R3_2R3_{1\alpha})/R_3+R3_\alpha(\cos\alpha+R3_1R3_2)/R_3^2$$

Quartic

$$R3_{1111}=-2(R3_1R3_{111}+R3_{11}^2)/R_3-R3_{11}(1+R3_1^2)/R_3^2+2R3_1^2(1+R3_1^2)/R_3^3$$
$$R3_{\alpha\alpha\alpha\alpha}=-R3_{\alpha\alpha}+(R3_\alpha^2-2R3_\alpha R3_{\alpha\alpha\alpha}-2R3_{\alpha\alpha}^2)/R_3+R3_{\alpha\alpha}(5R3_\alpha^2-R_1R_2\cos\alpha)/R_3^2$$
$$+2R3_\alpha^2(R_1R_2\cos\alpha-R3_\alpha^2)/R_3^3$$
$$R3_{1112}=-2(R3_1R3_{112}+R3_{11}R3_{12})/R_3+(2R3_1R3_2R3_{11}-R3_{12}$$
$$-3R3_{12}R3_1^2)/R_3^2+2R3_1R3_2(1+R3_1^2)/R_3^3$$
$$R3_{1122}=-2(R3_1R3_{122}+R3_{12}^2)/R_3-R3_{22}(1+R3_1^2)/R_3^2+2R3_2^2(1+R3_1^2)/R_3^3$$
$$R3_{111\alpha}=-2(R3_1R3_{11\alpha}+R3_{11}R3_{1\alpha})/R_3+(2R3_1R3_\alpha R3_{11}-R3_{1\alpha}$$
$$-3R3_1^2R3_{1\alpha})/R_3^2+2R3_1R3_\alpha(1+R3_1^2)/R_3^3$$
$$R3_{11\alpha\alpha}=-2(R3_1R3_{1\alpha\alpha}+R3_{1\alpha}^2)/R_3-R3_{\alpha\alpha}(1+R3_1^2)/R_3^2+2R3_\alpha^2(1+R3_1^2)/R_3^3$$
$$R3_{1\alpha\alpha\alpha}=-(R_2\sin\alpha+R3_1R3_{\alpha\alpha\alpha}+R3_\alpha R3_{1\alpha\alpha}+2R3_{1\alpha}R3_{\alpha\alpha})/R_3-(2R_2\cos\alpha R3_\alpha$$
$$+R_2\sin\alpha R3_{\alpha\alpha}-3R3_1R3_\alpha R3_{\alpha\alpha}-2R3_\alpha^2R3_{1\alpha})/R_3^2-2R3_\alpha^2(R_2\sin\alpha$$
$$-R3_1R3_\alpha)/R_3^3$$
$$R3_{112\alpha}=-2(R3_1R3_{12\alpha}+R3_{12}R3_{1\alpha})/R_3+(2R3_1R3_\alpha R3_{12}-R3_{2\alpha}-R3_1^2R3_{2\alpha}$$
$$-2R3_1R3_2R3_{1\alpha})/R_3^2+2R3_2R3_\alpha(1+R3_1^2)/R_3^3$$
$$R3_{12\alpha\alpha}=(\cos\alpha-R3_1R3_{2\alpha\alpha}-R3_2R3_{1\alpha\alpha}-2R3_{1\alpha}R3_{2\alpha})/R_3-(2\sin\alpha R3_\alpha$$
$$-2R3_1R3_\alpha R3_{2\alpha}-2R3_2R3_\alpha R3_{1\alpha}-\cos\alpha R3_{\alpha\alpha}-R3_1R3_2R3_{\alpha\alpha})/R_3^2$$
$$-2R3_\alpha^2(\cos\alpha+R3_1R3_2)/R_3^3$$

Appendix 1a

Special relationships for linear molecules

For linear molecules, $R_1 + R_2 = R_3$, $\alpha = \pi$. The expressions for the force constants take the following special forms.

Linear

$$W_1 = V_1 + V_3$$
$$W_\alpha = 0$$

Quadratic

$$W_{11} = V_{11} + 2V_{13} + V_{33}$$
$$W_{\alpha\alpha} = V_3 R3_{\alpha\alpha}$$
$$W_{12} = V_{12} + V_{13} + V_{23} + V_{33}$$
$$W_{1\alpha} = 0$$

Cubic

$$W_{111} = V_{111} + 3V_{113} + 3V_{133} + V_{333} + V_3 R3_{111}$$
$$W_{\alpha\alpha\alpha} = 0$$
$$W_{112} = V_{112} + V_{113} + 2V_{123} + 2V_{133} + V_{233} + V_{333} + V_3 R3_{112}$$
$$W_{11\alpha} = 0$$
$$W_{1\alpha\alpha} = V_{13} R3_{\alpha\alpha} + V_{33} R3_{\alpha\alpha} + V_3 R3_{1\alpha\alpha}$$
$$W_{12\alpha} = 0$$

Quartic

$$W_{1111} = V_{1111} + 4V_{1113} + 6V_{1133} + 4V_{1333} + V_{3333} + 4V_{13} R3_{111} + 4V_{33} R3_{111} + V_3 R3_{1111}$$
$$W_{\alpha\alpha\alpha\alpha} = 3V_{33} R3^2_{\alpha\alpha} + V_3 R3_{\alpha\alpha\alpha\alpha}$$
$$W_{1112} = V_{1112} + V_{1113} + 3V_{1123} + 3V_{1133} + 3V_{1233} + 3V_{1333} + V_{2333} + V_{3333} + 3V_{13} R3_{112} + V_{23} R3_{111} + V_{33}(3R3_{112} + R3_{111}) + V_3 R3_{1112}$$
$$W_{1122} = V_{1122} + 2V_{1123} + 2V_{1223} + V_{1133} + V_{2233} + 4V_{1233} + 2V_{1333} + 2V_{2333} + V_{3333} + 2V_{13} R3_{122} + 2V_{23} R3_{112} + 2V_{33}(R3_{122} + R3_{112}) + V_3 R3_{1122}$$
$$W_{111\alpha} = 0$$
$$W_{11\alpha\alpha} = V_{113} R3_{\alpha\alpha} + 2V_{133} R3_{\alpha\alpha} + V_{333} R3_{\alpha\alpha} + 2V_{13} R3_{1\alpha\alpha} + 2V_{33} R3_{1\alpha\alpha} + V_3 R3_{11\alpha\alpha}$$

$$W_{1\alpha\alpha\alpha}=0$$
$$W_{112\alpha}=0$$
$$W_{12\alpha\alpha}=V_{333}R3_{\alpha\alpha}+V_{133}R3_{\alpha\alpha}+V_{233}R3_{\alpha\alpha}+V_{123}R3_{\alpha\alpha}+V_{33}(R3_{1\alpha\alpha}+R3_{2\alpha\alpha})+V_{13}R3_{2\alpha\alpha}+V_{23}R3_{1\alpha\alpha}+V_3R3_{12\alpha\alpha}$$

where the geometrical relationships are given by the following.

Linear

$$R3_1=1$$
$$R3_\alpha=0$$

Quadratic

$$R3_{11}=0$$
$$R3_{\alpha\alpha}=-R_1R_2/R_3$$
$$R3_{12}=0$$
$$R3_{1\alpha}=0$$

Cubic

$$R3_{111}=-2/R_3^2$$
$$R3_{\alpha\alpha\alpha}=0$$
$$R3_{112}=-2/R_3^2$$
$$R3_{11\alpha}=0$$
$$R3_{1\alpha\alpha}=-(R_2/R_3)^2$$
$$R3_{12\alpha}=0$$

Quartic

$$R3_{1111}=8/R_3^3$$
$$R3_{\alpha\alpha\alpha\alpha}=R_1R_2(1-3R_1R_2/R_3^2)/R_3$$
$$R3_{1112}=8/R_3^3$$
$$R3_{1122}=8/R_3^3$$
$$R3_{111\alpha}=0$$
$$R3_{11\alpha\alpha}=2R_2/R_3^2$$
$$R3_{1\alpha\alpha\alpha}=0$$
$$R3_{112\alpha}=0$$
$$R3_{12\alpha\alpha}=-2R_1R_2/R_3^3$$

Appendix 2

Coordinate transformations

Let the triatomic molecule ABC be defined by bond distance coordinates R_i, $i=1-3$. We define optimized displacement coordinates

$$\boldsymbol{S}=\boldsymbol{\Gamma}\boldsymbol{\rho}$$

in terms of the translation $\boldsymbol{\rho}=\boldsymbol{R}-\boldsymbol{R}^0$ and the rotation $\boldsymbol{\Gamma}(\alpha, \beta, \gamma)$. The translation fixes the reference structure R_i^0 of the bond displacements ρ_i. The rotation matrix $\boldsymbol{\Gamma}(\alpha, \beta, \gamma)$ defines rotations of ρ_i through the Euler angles α, β, γ:

$$\boldsymbol{\Gamma}(\alpha, \beta, \gamma)=\Gamma_3(\gamma) \quad \Gamma_2(\beta) \quad \Gamma_1(\alpha)$$

$$\Gamma_1(\alpha)=\begin{bmatrix} \cos\alpha & 0 & -\sin\alpha \\ 0 & 1 & 0 \\ \sin\alpha & 0 & \cos\alpha \end{bmatrix} \qquad \text{(rotation about } \rho_2\text{)}$$

$$\Gamma_2(\beta)=\begin{bmatrix} 1 & 0 & 0 \\ 0 & \cos\beta & \sin\beta \\ 0 & -\sin\beta & \cos\beta \end{bmatrix} \qquad \text{(rotation about } \rho_1'\text{)}$$

$$\Gamma_3(\gamma)=\begin{bmatrix} \cos\gamma & 0 & -\sin\gamma \\ 0 & 1 & 0 \\ \sin\gamma & 0 & \cos\gamma \end{bmatrix} \qquad \text{(rotation about } \rho_2''\text{)}$$

The operation of Γ_1, Γ_2 and Γ_3 successively on ρ_i produces ρ_i', ρ_i'' and $\rho_i'''=S_i$,

The translation and rotation parameters which define S_i are given in Table A2.1 for systems containing one, two or three minima on the surface and which may or may not possess symmetry. For case (i), the single minimum is located at $(S_1, S_2, S_3)=(0, 0, 0)$. For cases (ii) and (iii) the two minima are located at $(0, \pm S_{2a}, 0)$. For case (iv), the S_i coordinates are defined by symmetry (see Section 4.2), the three minima being located at $(0, 0, S_{3a})$ and $(0, \pm S_{2b}, -\frac{1}{2}S_{3a})$. For case (v), two minima are located at $(0, \pm S_{2a}, 0)$, with the third minimum at $(0, 0, S_{3c})$. For case (vi), minima *a* and *b* are located at $(0, S_{2a}, 0)$ and $(0, S_{2b}, 0)$, respectively, and minimum *c* is located at $(0, 0, S_{3c})$.

Note that for the simplified two-dimensional problems discussed in Sections 4.2 and 4.3, R_1 and R_2 were taken as the equivalent bonds in an AB_2 system, so

Table A2.1 Expressions which define the optimized coordinates S_i in terms of the internuclear distances R_i, according to a translation $\rho_i = R_i - R_i^0$, and a rotation $\boldsymbol{S} = \boldsymbol{\Gamma}\boldsymbol{\rho}$.

	Translation	Rotation
(i) 1 minimum	$R_i^0 = R_{ia}, \quad i = 1, 3$	$\alpha = \beta = \gamma = 0$
(ii) 2 minima 2-fold symmetry	$R_1^0 = R_3^0 = (R_{1a} + R_{1b})/2$ $R_2^0 = R_{2b}$	$\alpha = 3\pi/4$ $\beta = \pi/2$ γ undefined*
(iii) 2 minima no symmetry	$R_i^0 = (R_{ia} + R_{ib})/2, \quad i = 1, 3$	$\alpha = \tan^{-1}(\rho_{1a}/\rho_{3a})$† $\beta = \tan^{-1}(\rho'_{3a}/\rho'_{2a})$ γ undefined*
(iv) 3 minima 3-fold symmetry	$R_i^0 = (R_{ia} + R_{ib} + R_{ic})/3, \quad i = 1, 3$	$\alpha = 3\pi/4$ $\beta = \pi/2$ $\gamma = 0.8041\pi$
(v) 3 minima 2-fold symmetry $(a = b)$	$R_1^0 = R_3^0 = (R_{1a} + R_{1b})/2$ $R_2^0 = R_{2a} = R_{2b}$	$\alpha = 3\pi/4$ $\beta = \pi/2$ $\gamma = \tan^{-1}(\rho''_{1c}/\rho''_{3c})$
(vi) 3 minima no symmetry	$R_i^0 = (1 - k)R_{ia} + kR_{ib}$ $k = \dfrac{\sum_{i=1}^{3} (R_{ic} - R_{ia})(R_{ib} - R_{ia})}{\sum_{i=1}^{3} (R_{ib} - R_{ia})^2}$	$\alpha = \tan^{-1}(\rho_{1a}/\rho_{3a})$ $\beta = \tan^{-1}(\rho'_{3a}/\rho'_{2a})$ $\gamma = \tan^{-1}(\rho''_{1c}/\rho''_{3c})$

* Although the S_i coordinates are optimized to the geometries of a and b for all values of γ, there may be other conditions for which a best or convenient value of γ can be found. For example, $\gamma = 0.8041\pi$, as in case (iv), is often convenient.

† For all angles specified by an expression $\tan^{-1}(f/g)$, the sine of the angle has the sign of f.

Table A2.2 Coefficients in $P(S_1, S_2, S_3)$ which are determined from information relating to three minima a, b and c on the potential energy surface for which no symmetry relationship exists between any of the three minima. Minimum a $(0, S_{2a}, 0)$ determines all the coefficients up to quadratic.

Data	Terms from $b(0, S_{2b}, 0)$	Terms from $c(0, 0, S_{3c})$
Energy	S_2^3	S_3^3
$\partial/\partial S_1$	$S_1 S_2^2$	$S_1 S_3^2$
$\partial/\partial S_2$	S_2^4	$S_2 S_3^2$
$\partial/\partial S_3$	$S_2^2 S_3$	S_3^4
$\partial^2/\partial S_1^2$	$S_1^2 S_2$	$S_1^2 S_3$
$\partial^2/\partial S_2^2$	S_2^5	$S_2^2 S_3^2$
$\partial^2/\partial S_3^2$	$S_2 S_3^2$	S_3^5
$\partial^2/\partial S_1 \partial S_2$	$S_1 S_2^2$	$S_1 S_2 S_3^2$
$\partial^2/\partial S_1 \partial S_3$	$S_1 S_2 S_3$	$S_1 S_3^3$
$\partial^2/\partial S_2 \partial S_3$	$S_2^3 S_3$	$S_2 S_3^3$

Table A2.3 Coefficients in $P(S_1, S_2, S_3)$ which are determined for an X_2Y molecule. The table assumes data are complete for minima *a* and *b* and complete or incomplete for *c*. In the case where *c* is complete but *a* and *b* are known only in respect of energy and geometry, the terms are shown in parentheses.

Data	Terms from $a(0, S_{2a}, 0)$ and $b(0, -S_{2a}, 0)$		Terms from $c(0, 0, S_{3c})$	
Energy	Constant	(S_2^2)	S_3^3	(Constant)
$\partial/\partial S_1$	S_1	(S_1)	$S_1S_3^2$	(S_1S_3)
$\partial/\partial S_2$	S_2^2	(S_2^4)	—	—
$\partial/\partial S_3$	S_3	(S_3)	S_3^4	(S_3^2)
$\partial^2/\partial S_1^2$	S_1^2		$S_1^2S_3$	(S_1^2)
$\partial^2/\partial S_2^2$	S_2^4		$S_2^2S_3^2$	$(S_2^2S_3)$
$\partial^2/\partial S_3^2$	S_3^2		S_3^5	(S_3^3)
$\partial^2/\partial S_1\partial S_2$	$S_1S_2^2$		—	—
$\partial^2/\partial S_1\partial S_3$	S_1S_3		$S_1S_3^3$	$(S_1S_3^2)$
$\partial^2/\partial S_2\partial S_3$	$S_2^2S_3$		—	—

the expressions used there are not directly derived from the more general expressions in Table A2.1.

Tables A2.2 and A2.3 list the various coefficients of the polynomial $P(S_i)$ which can be determined from the spectroscopic data for minima *a*, *b* and *c*, where the S_i coordinates have been derived from the parameters given in Table A3.1. It must be emphasized that the terms chosen are the lowest order that can be determined from a particular piece of data, and that higher-order terms could be substituted. For example, from the data pertaining to minimum *c* in Table A2.2, higher powers of S_3 could be used because the variable S_3 is non-zero at this minimum.

Appendix 3

Equations relating derivatives of the potential and coefficients of the three-body polynomial

Let the reference structure R_i^0 defining the coordinates $\rho_i = R_i - R_i^0$ coincide with the origin R_i^e of the Taylor expansion of the potential and let C_{ij}, for example, be the coefficient of $\rho_i \rho_i$ in the polynomial $P(R_1, R_2, R_3)$ of expression (7.5). If

$$G_{ij} = (V^0)^{-1} (\partial^2 V^{(3)}_{\mathrm{ABC}}(R_1, R_2, R_3)/\partial R_i\, \partial R_j)_e$$

where $V^{(3)}_{\mathrm{ABC}}(R_1, R_2, R_3)$ is given by equation (7.3), V^0 is given by equation (7.7) and

$$g_i = \gamma_i/2$$

where γ_i is defined in equation (7.6), then the relevant equations are as follows:

$$
\begin{aligned}
C_i &= g_i + G_i \\
C_{ii} &= g_i^2 + g_i G_i + G_{ii}/2 \\
C_{ij} &= g_i g_j + g_i G_j + g_j G_i + G_{ij} \\
C_{iii} &= 2g_i^3/3 + g_i^2 G_i + g_i G_{ii}/2 + G_{iii}/6 \\
C_{iij} &= g_i^2 g_j + g_i g_j G_i + g_i^2 G_j + g_i G_{ij} + g_j G_{ii}/2 + G_{iij}/2 \\
C_{ijk} &= g_i g_j g_k + g_i g_j G_k + g_i g_k G_j + g_i g_k G_i + g_i G_{jk} + g_j G_{ik} + g_k G_{ij} + G_{ijk} \\
C_{iiii} &= g_i^4/3 + 2g_i^3 G_i/3 + g_i^2 G_{ii}/2 + g_i G_{iii}/6 + G_{iiii}/24 \\
C_{iiij} &= 2g_i^3 g_j/3 + g_i^2 g_j G_i + 2g_i^3 G_j/3 + g_i g_j G_{ii}/2 + g_i^2 G_{ij} + g_j G_{iii}/6 \\
&\quad + g_i G_{iij}/2 + G_{iiij}/6 \\
C_{iijj} &= g_i^2 g_j^2 + g_i^2 g_j G_j + g_i g_j^2 G_i + g_i g_j G_{ij} + g_i^2 G_{jj}/2 + g_j^2 G_{ii}/2 \\
&\quad + g_j G_{iij}/2 + g_i G_{ijj}/2 + G_{iijj}/4 \\
C_{iijk} &= g_i^2 g_j g_k + g_i g_j g_k G_i + g_i^2 g_k G_j + g_i^2 g_j G_k + g_i g_j G_{ik} + g_i g_k G_{ij} \\
&\quad + g_j g_k G_{ii}/2 + g_i^2 G_{jk} + g_j G_{iik}/2 + g_k G_{iij}/2 + g_i G_{ijk} + G_{iijk}/2
\end{aligned}
$$

Subject Index

Index of triatomic and tetra-atomic systems